THE BEDFORD SERIES IN HISTORY AND CULTURE

Women's Rights Emerges within the Antislavery Movement, 1830–1870

A Brief History with Documents

Related Titles in

THE BEDFORD SERIES IN HISTORY AND CULTURE

Judith Sargent Murray: A Brief Biography with Documents
Sheila L. Skemp, *University of Mississippi*

Narrative of the Life of Frederick Douglass, An American Slave, Written by Himself
Edited with an Introduction by David W. Blight, *Amherst College*

Margaret Fuller: A Brief Biography with Documents
Eve Kornfeld, *San Diego State University*

William Lloyd Garrison and the Fight against Slavery: Selections from THE LIBERATOR
Edited with an Introduction by William E. Cain, *Wellesley College*

Defending the Cornerstone: Proslavery Arguments in the American South (forthcoming)
Paul Finkelman, *University of Tulsa College of Law*

Dred Scott v. Sandford: A Brief History with Documents
Paul Finkelman, *University of Tulsa College of Law*

Abraham Lincoln, Slavery, and the Civil War: Selected Writings and Speeches (forthcoming)
Michael P. Johnson, *Johns Hopkins University*

African American Perspectives on the Civil War and Reconstruction (forthcoming)
Michael P. Johnson, *Johns Hopkins University*

Southern Horrors and Other Writings: The Anti-Lynching Campaign of Ida B. Wells, 1892–1900
Edited with an Introduction by Jacqueline Jones Royster, *The Ohio State University*

THE BEDFORD SERIES IN HISTORY AND CULTURE

Women's Rights Emerges within the Antislavery Movement, 1830–1870

A Brief History with Documents

Kathryn Kish Sklar

State University of New York, Binghamton

BEDFORD/ST. MARTIN'S Boston ♦ New York

For Bedford/St. Martin's
Executive Editor for History and Political Science: Katherine E. Kurzman
Developmental Editor: Louise Townsend
Editorial Assistant: Chip Turner
Senior Production Supervisor: Dennis Conroy
Marketing Manager: Charles Cavaliere
Project Management: Books By Design, Inc.
Text Design: Claire Seng-Niemoeller
Indexer: Kathleen Babbitt
Cover Design: Richard Emery Design, Inc.
Cover Art: Am I Not a Woman and a Sister? Logo for the Ladies Department of *The Liberator.* By permission of the Houghton Library, Harvard University.
Composition: G&S Typesetters, Inc.
Printing and Binding: Haddon Craftsmen, an R. R. Donnelley & Sons Company

President: Charles H. Christensen
Editorial Director: Joan E. Feinberg
Director of Marketing: Karen R. Melton
Director of Editing, Design, and Production: Marcia Cohen
Manager, Publishing Services: Emily Berleth

Library of Congress Catalog Card Number: 99-63693

Softcover reprint of the hardcover 1st edition 2000 978-0-312-22819-4

Manufactured in the United States of America.

5 4 3 2 1 0
f e d c b a

For information, write: Bedford/St. Martin's, 75 Arlington Street, Boston, MA 02116
(617-399-4000)

ISBN: 978-0-312-10144-2 (paperback)
ISBN 978-1-349-62638-0 ISBN 978-1-137-04527-0 (eBook)
DOI 10.1007/978-1-137-04527-0

TO MY DAUGHTERS

Susan Sklar Friedman
Amy Luers
Sascha Dublin
Sonya Dublin

Wondrous new progenitors of women's rights

Foreword

The Bedford Series in History and Culture is designed so that readers can study the past as historians do.

The historian's first task is finding the evidence. Documents, letters, memoirs, interviews, pictures, movies, novels, or poems can provide facts and clues. Then the historian questions and compares the sources. There is more to do than in a courtroom, for hearsay evidence is welcome, and the historian is usually looking for answers beyond act and motive. Different views of an event may be as important as a single verdict. How a story is told may yield as much information as what it says.

Along the way the historian seeks help from other historians and perhaps from specialists in other disciplines. Finally, it is time to write, to decide on an interpretation and how to arrange the evidence for readers.

Each book in this series contains an important historical document or group of documents, each document a witness from the past and open to interpretation in different ways. The documents are combined with some element of historical narrative — an introduction or a biographical essay, for example — that provides students with an analysis of the primary source material and important background information about the world in which it was produced.

Each book in the series focuses on a specific topic within a specific historical period. Each provides a basis for lively thought and discussion about several aspects of the topic and the historian's role. Each is short enough (and inexpensive enough) to be a reasonable one-week assignment in a college course. Whether as classroom or personal reading, each book in the series provides firsthand experience of the challenge — and fun — of discovering, recreating, and interpreting the past.

Natalie Zemon Davis
Ernest R. May

"I am persuaded that woman is not to be as she has been, a mere second-hand agent in the regeneration of a fallen world, but the acknowledged equal and co-worker with man in this glorious work."

ANGELINA GRIMKÉ
in a letter written from Groton, Mass., August 10, 1837

Preface

The first women's rights movement in the United States, which began in the 1830s, emerged from the campaign to end slavery. As active participants in the effort to abolish slavery, white women and free black women mobilized their communities, created local organizations, and provided crucial financial support to the national antislavery association. These efforts were so successful that women became unwilling to abide by the rules that limited their participation in public life. In defending themselves against critics who deplored their departure from "woman's assigned sphere," they launched a women's rights agenda that resonates to the present day.

The chief actors in this dramatic story were Angelina and Sarah Grimké, daughters of a prominent slaveholding family in Charleston, South Carolina. Their migration out of the South and their recruitment into the most radical form of antislavery protest set the stage for their innovations on behalf of women's rights. In the process of breaking down the barriers of opinion that prohibited women from speaking in public, they created a platform of women's rights that insisted "whatever is morally right for a man to do is morally right for a woman to do."

This book focuses on the process by which the Grimké sisters and their supporters transformed the definition of acceptable behavior for women in public life in the United States. It gathers selections from their letters and writings as well as from others who illuminate the context within which they acted in the 1830s. The book also considers how an independent women's rights movement came into being in the middle decades of the nineteenth century. Using the Grimkés' legacy and responding to changes in American society and culture, that movement reveals a great deal about American life in the decades before and the years immediately after the Civil War.

Taking a fresh look at the documents associated with these remarkable events, this book offers a fresh interpretation of them — one that emphasizes the religious context of the Grimkés' protest and their commitment to racial equality as well as gender equality. Also new is the book's account of how the women's rights movement was transplanted into more secular soil after 1840.

I have chosen documents that reveal the subjective experience of the early supporters of women's rights. For them women's rights was more than an ideology. It reflected their immediate experience, both public and private. This subjectivity was an important basis for the Grimkés' innovations, and it became a major reason for the movement's rapid growth after 1840.

This selection of documents also shows how questions about race and racial prejudice shaped the women's rights movement from the moment of its birth in 1837 to its institutionalization within the first national women's suffrage associations in 1869. This book is a study in the history of race as well as the history of gender.

These documents were also chosen with an eye to illuminating the dynamic impact of the antislavery movement on American society in the 1830s. Of incalculable significance in U.S. history, the abolitionist movement remains one of the great reservoirs of civil morality in American political culture.

I also selected documents that help us understand how civil society — that is, groups active outside the institutions of government — were changing during a crucial period of American history. The culture of public life is a crucial ingredient in any society, never more so than on the eve of the American Civil War. The rapidly changing place of women within that culture provides a wide window for viewing public life.

Because social movements are deeply embedded in the culture that produces them, to understand them we must explore their cultural contexts. This book examines the cultural environment of the antebellum women's rights movement by looking at the way that public culture was constructed with values drawn from ideologies about gender, race, religion, region, and family life.

These documents also invite us to witness how history has been shaped by the contingent, the unexpected, and the unlikely. No general laws could have forecast the events depicted in these pages. There is a love story here — but even it has an unpredictable outcome.

ACKNOWLEDGMENTS

All historians, especially those who edit primary sources, are indebted to archivists who aid in their quest for the perfect document. I am particularly grateful to William Faucon in the Rare Books and Manuscripts Department of the Boston Public Library, and Arlene Shy at the William L. Clements Library in Early American History at the University of Michigan.

I am also indebted to the scholars whose work on the Grimké sisters and on the early women's rights movement has provided me with such an abundance of insightful historical analysis. Gerda Lerner and Katharine Lumpkin have set high standards that I have tried to follow in my research on the Grimkés. I am enormously grateful to Larry Ceplair's pathbreaking collection of documents on the Grimkés' public years, which provides me and other scholars in the field with a definitive edition on the subject. Shirley Yee's work on black women abolitionists, Julie Roy Jeffrey's book on ordinary women in the antislavery movement, Dorothy Sterling's study of Abby Kelley, and Nancy Hewitt's analysis of women's activism in antebellum Rochester have all contributed greatly to this book, as have classic volumes on antislavery by Robert Abzug and Lewis Perry.

I am grateful to the folks at Bedford/St. Martin's for their author-friendly approach to publishing. Katherine E. Kurzman offered steady encouragement. Louise Townsend's helpful editorial hand gently but firmly moved the book along to publication. Chip Turner unfailingly supplied the book's many needs as it moved through Bedford/St. Martin's. Joan Flaherty provided excellent copyediting, Halle Lewis read the page proofs with remarkable care, and Melissa Lotfy of Books By Design guided the book to completion.

Discussions with Carla Peterson helped me see the importance of Frances Watkins Harper in the 1860s debates. Rifa'at Abou El-Haj pointed out the crucial significance of Angelina Grimké's moratorium in the early 1830s. I am grateful to John McClymer for the encouragement of his fine example in his edited collection of documents from the Worcester Woman's Rights Convention of 1850, and his help in locating documents that prove particularly pertinent to my interpretation here.

Readers of the manuscript were very helpful. Robert Abzug, Ruth Alexander, Nancy Hewitt, Julie Roy Jeffrey, and John McClymer improved the book by generously sharing their rich knowledge of the topic. Tom Dublin read the manuscript in almost all its incarnations, always offering valuable insights and timely encouragements.

Kathryn Kish Sklar

EDITORIAL NOTE

To achieve greater clarity for the modern reader, some spellings and punctuations have been modernized, minor errors silently corrected, and some paragraph breaks added. Abbreviations have been retained, however. Most Quaker usages have been retained, but some Quaker dates have been modernized. All italics appear in the originals; none were added.

Contents

PART TWO

The Documents 77

APPENDICES

Illustrations

PART ONE

Introduction: "Our Rights as Moral Beings"

In the summer of 1837, two sisters from South Carolina, Angelina and Sarah Grimké, age 32 and 45 respectively, began a speaking tour of New England that permanently altered American perceptions of the rights of women. What began as a tour to promote the abolition of slavery ended by introducing the new concept of women's rights into American public life.[1] Between May and September the Grimkés ignited a debate about the equality of the sexes that first enveloped the abolitionist movement and then extended into the lives of women who were active in other reforms, precipitating large changes in consciousness in a relatively short period of time.

Although the Grimké sisters retired from public life in 1839, their ideas took root in the fertile soil of a rapidly changing society. In 1848 a Women's Rights Convention at Seneca Falls, New York, marked the emergence of women's rights as an autonomous social movement. A cascade of women's rights conventions in the 1850s carried the movement into towns and villages throughout the Northeast and Midwest. After the Civil War the movement consolidated around women's right to vote,

[1] Although the term "woman's rights" was used more frequently in the nineteenth century than "women's rights," I use the term "women's rights" because it has become the preferred term in the twentieth century.

forming two organizations in 1869—the National Woman Suffrage Association in New York and the American Woman Suffrage Association in Boston. For the next fifty years, until the ratification of the Nineteenth Amendment to the U.S. Constitution granted votes to women in 1920, the woman suffrage movement forged channels through which women contributed enormous vitality and energy to American civil society.

The emergence of an autonomous women's rights movement from the struggle against slavery was not inevitable. Although women were also active in British antislavery circles, for example, their activism did not generate an equivalent women's rights movement in England. In the United States a movement arose out of the confluence of many causes, some rooted in the particulars of American slavery, American religious culture, and American political culture, some arising from other changes already under way in the personal lives of American women between 1830 and 1870. The movement began, however, with the unique position the Grimké sisters, exiles from South Carolina, held within the campaign to end slavery.

PRELUDE: BREAKING AWAY FROM SLAVE SOCIETY

Angelina and Sarah Grimké occupied a special position in the antislavery movement between 1835 and 1839 because they were the daughters of an elite southern slaveholding family who left the South and became lecturers for the American Anti-Slavery Society. Their compelling descriptions of the horrors of slavery attracted audiences that otherwise might have remained indifferent to the topic, given the wide extent to which livelihoods in the North were intertwined with the economic success of slavery in the South. For example, the politically powerful class of northern merchants profited substantially by selling southern cotton to English manufacturers. And even New England textile workers depended on southern cotton for their earnings.[2]

Before 1830 the Grimké sisters seemed an unlikely pair to launch a revolution. Born to a prominent slave-owning family of Huguenot (French Protestant) descent in Charleston, South Carolina, Sarah was the sixth and Angelina the last of fourteen children. Their father was

[2] John Ashworth, *Slavery, Capitalism, and Politics in the Antebellum Republic, Commerce and Compromise, 1820–1850* (New York: Cambridge University Press, 1995), 1:365.

chief justice of the state's highest court; their mother's brother served as governor of North Carolina. The family's wealth derived from a plantation that they never visited, two hundred miles away in the Carolina "upcountry," which was run by a hired overseer. Their substantial house in Charleston required the labor of many slaves, most of whom had relatives on the plantation.[3]

The seeds for the sisters' abhorrence of slavery were sown within its embrace; during their privileged childhood they sympathized with the sufferings of slave children. For example, they observed the forced separation of slave parents and children. Sometimes the children of house servants were sent to the plantation; sometimes children were brought from the plantation to work in the house. In either case, children's bonds with their parents were ignored and they were treated like property rather than human beings.

A girlhood friend of Angelina lived near the place where slaves were sent to be punished, and "often heard the screams of the slaves under their torture." At about age thirteen Angelina attended a school where "nearly all the aristocracy" sent their daughters. One day a young slave boy about her age was called into the classroom to open a window. He had been so "dreadfully whipped that he could hardly walk." His "heart-broken countenance" awed her as much as his bloodied back.[4]

Sarah and Angelina's mother, Mary Grimké, supervised her household with callous disregard for the well-being of the slaves who served her. Bondspeople slept on the floor without bed or bedding, ate from tin plates without a table, and had no lights in their quarters. Seamstresses were required to work after dark in winter by staircase lamps, which were so dim that they had to stand to see their work. "Mother," Angelina wrote in her diary in 1828, "rules slaves and children with a rod of fear."[5] After their father died in 1819, and brother Henry became the head of household, slave punishments became even more severe. One of

[3]A brief biography of Angelina and Sarah Grimké (hereafter AG and SG) and most other women mentioned in this essay can be found in Edward James et al., eds., *Notable American Women: A Biographical Dictionary* (Cambridge: Harvard University Press, 1971). See also footnote 8.

[4]"Testimony of Angelina Grimké Weld," in *American Slavery as It Is: Testimony of a Thousand Witnesses,* Theodore D. Weld, ed. (New York: American Anti-Slavery Society, 1839), 54–55.

[5]Large portions from AG's diary have been printed in Catherine H. Birney, *The Grimké Sisters: Sarah and Angelina Grimké: The First American Women Advocates of Abolition and Women's Rights* (Philadelphia: Lee and Shepard, 1885), 55–93. The treatment of slaves in the Grimké household can be inferred from passages in "Testimony of Angelina Grimké Weld" and "Narrative and Testimony of Sarah M. Grimké," in *American Slavery as It Is,* 52–57 and 22–24.

Henry's blows to the head of the family butler left the servant subject to epileptic fits. Disorder was a close companion of violence in the Grimké household. When her sisters blamed the family's slaves for the household's lack of order and discipline, Angelina replied that "the servants were just what the family was . . . not at all more rude and selfish and disobliging . . . and how could they expect the servants to behave, when they had such examples continually before them?"[6]

One important subtext of the cruelty within slave-owning households was, of course, the fact that slaves were held in bondage against their will. Charleston's twenty thousand whites in 1820 were far outnumbered by the city's sixty thousand slaves, a ratio that bred fears of slave revolts and brutal reprisals against potential uprisings. In Charleston in the summer of 1822 a free black carpenter, Denmark Vesey, secretly organized an extensive slave revolt that was discovered before it erupted. Vesey and thirty-six other black men were hanged.[7]

Yet like most southern critics of slavery, Angelina and Sarah Grimké would almost certainly have accepted their circumstances as beyond their ability to change had they not embarked on a search for a more meaningful religious faith than the comfortable Episcopalianism of their mother. Each in her own way came to reject slavery by first rejecting the religious alternatives available in Charleston and migrating to join the Quaker community in Philadelphia, where each came to accept the Quaker view of slavery as sinful.[8]

Sarah, thirteen years older than Angelina, went first. Touched by the powerful forces of evangelical Protestantism and the tremendous groundswell of religious fervor known as "the Second Great Awakening," Sarah had experienced religious conversion in Charleston under the guidance of a visiting Presbyterian minister in 1813. He preached that her soul would never be saved while she enjoyed fashionable

[6]AG, Diary, June 12, 1828. Weld-Grimké Collection, William L. Clements Library, University of Michigan, Ann Arbor.

[7]See John Lofton, *Denmark Vesey's Revolt: The Slave Plot That Lit a Fuse to Fort Sumter* (Kent, Ohio: Kent State University Press, 1983); and Edward A. Pearson, ed., *Designs against Charleston: The Trial Record of the Denmark Vesey Slave Conspiracy of 1822* (Chapel Hill: University of North Carolina Press, 1999). The slave population in the United States at this time was about two million.

[8]This interpretation of the Grimké sisters draws on my research in primary materials. It builds on but is not the same as interpretations found in Birney, *The Grimké Sisters;* Larry Ceplair, ed., *The Public Years of Sarah and Angelina Grimké: Selected Writings, 1835–1839* (New York: Columbia University Press, 1989); Gerda Lerner, *The Grimké Sisters of North Carolina: Pioneers for Women's Rights and Abolitionism* (New York: Oxford University Press, 1967); Gerda Lerner, *The Feminist Thought of Sarah Grimké* (New York: Oxford University Press, 1998); and Katharine DePre Lumpkin, *The Emancipation of Angelina Grimké* (Chapel Hill: University of North Carolina Press, 1974).

Charleston society — dances, teas, house parties, and other frivolities. This conversion was short-lived but it instilled long-lasting guilt over her privileged life and severed her from self-satisfied Episcopalianism. In 1819 when Sarah traveled with her ailing father to consult Quaker physicians in Philadelphia and he died at a seaside resort they had recommended, she was befriended by Quakers. After his death she lingered two months under their calming influence, absorbing their spirituality, belief in the unmediated relationship between God and the individual conscience, plain style of dress, few servants, orderly habits, and condemnation of slavery as ungodly.[9] After two years back in Charleston, where she found nothing that could match the appeal of the Philadelphia Quaker community, Sarah returned to Philadelphia and soon thereafter converted to Orthodox Quakerism.

Sarah's ties with her family were stretched but not broken. Far from demonizing the North, her family, like those of many wealthy southerners, frequently drew on its resources. For example, her older brother, Thomas, to whom she had been closely attached as a child, graduated from Yale University in Connecticut. Sarah visited Charleston several times, including a stay of six months during the winter of 1827. There she revived the close relationship she had forged with Angelina since the time of her younger sister's birth. Sarah was Angelina's godmother and in many ways became a more meaningful parent than their mother. Angelina's letters to Sarah in Philadelphia addressed her as "dearest Mother" and Sarah's called her "my precious child."[10]

In 1827 Sarah returned to find Angelina at the center of a vibrant religious circle created by a young, northern-born Presbyterian minister and his congregation. Attracted by his successful community organizing, Angelina had formed an intensely personal relationship with the minister. At the church she led a "colored Sunday school," organized prayer meetings, participated in religious societies, did charity work among the poor, and attended a "female prayer meeting" where Baptists,

[9] Quakers led the way in the abolition of slavery in the North. In 1758 the Philadelphia Yearly Meeting of the Society of Friends voted to exclude anyone who bought or sold slaves; in 1776 it excluded anyone who owned slaves. Between 1780 and 1800 most Northern states enacted statutes that abolished slavery, though this was usually accomplished gradually. This was the case in the Pennsylvania "Act for the Gradual Abolition of Slavery" of 1780, which immediately emancipated all children born after the passage of the act, but emancipated adults more gradually. See "Abolition Statutes," in ***Children and Youth in America: A Documentary History, 1: 1600–1865,*** Robert H. Bremner et al., eds. (Cambridge: Harvard University Press, 1970), 324–26. In 1830 Negro slavery was still not totally abolished in the North; about 3,500 persons still remained in bondage, mostly in New Jersey. See Leon F. Litwack, *North of Slavery: The Negro in the Free States, 1790–1860* (Chicago: University of Chicago Press, 1961), 12–14.

[10] Lumpkin, *The Emancipation of Angelina Grimké,* 20.

Methodists, Congregationalists, and Presbyterians met monthly to discuss their responsibilities as women.[11] At home Angelina had organized daily prayer meetings that were attended by her mother and sisters, the family slaves, and slaves from other households. Influenced by the Second Great Awakening, many southern families and churches participated in such interracial efforts before 1830.[12]

A struggle for Angelina's soul ensued, with her "beloved" Presbyterian minister on one side and Sarah's Quakerism on the other. At issue was the question of whether Quakers or Presbyterians offered the more reliable route to religious salvation. Sarah won, but only after an intense battle. The older sister's victory was sealed when Angelina cut up her Walter Scott novels and stuffed a cushion with the laces, veils, and trimmings that had adorned her clothing. Three months after Sarah returned to Philadelphia, Angelina joined her there in July 1828.

SEEKING A VOICE: GARRISONIAN ABOLITIONIST WOMEN, 1831–1833

Years of uncertainty followed. The sisters' escape from Charleston had religious rather than antislavery motives.[13] They joined the conservative Quaker community that had taken Sarah in when their father died. That community, more interested in maintaining its internal hierarchy than in critiquing social injustice, was the antithesis of the one Angelina had recently relinquished. Her Charleston flock had been gregarious, stylish, and hospitable. Her new congregation valued silence, was deliberately dour, and prohibited mingling with outsiders. Participation in "popular" causes — that is, those not exclusively Quaker — was strongly discouraged. The sisters lived comfortably in the home of a Quaker woman who had befriended Sarah in 1819, and their inheritance from their father made it unnecessary for them to work for a living, but both sought greater meaning and usefulness. They decided to become Quaker ministers.

[11] AG to Elizabeth Bascom, July 23, 1828; and AG to (her sister) Mrs. Anna Frost, March 17, 1828, both in the Weld-Grimké Collection.

[12] Lerner, *The Grimké Sisters of North Carolina,* 54; Sylvia R. Frey and Betty Wood, *Come Shouting to Zion: African-American Protestantism in the American South and British Carribbean to 1830* (Chapel Hill: University of North Carolina Press, 1998), 182–208.

[13] Although AG later claimed that antislavery motivations impelled her departure from South Carolina (see Document 34), the historical evidence supports a more gradual evolution of her antislavery views.

Quakers did not support a "settled ministry"; instead their ministers were unpaid itinerants for whom the "inner light" of their calling sufficed to qualify them as clergy. Unlike other Protestants, Quakers sanctioned women ministers, though they expected women to minister primarily to the needs of their own sex. Separate women's meetings gave the Grimkés and other women the opportunity to gain speaking and leadership skills. Quaker religious meetings consisted of silence broken only by those (women as well as men) who were inspired, however briefly, to share their thoughts.[14]

Sarah took both her religious quest and her ministerial calling quite seriously. Although her speaking style was halting and tentative, she regularly expressed her thoughts in Sunday meeting, and traveled to visit and speak to other Quaker groups, especially groups of women. Angelina, by contrast, was less interested in the theological than the social practice of religion, never spoke in meeting, and grew increasingly alienated from the austere services. Searching for other avenues of usefulness, she explored the possibility of becoming a teacher, and visited Catharine Beecher's renowned female seminary in Hartford, Connecticut. Beecher, the daughter of prominent evangelical Congregational minister Lyman Beecher, was promoting the feminization of the teaching profession as an opportunity for women to become self-supporting. But Angelina's Quaker mentors disapproved because such a move would take her out of the community. Permitted by the congregation to return to Charleston to care for her mother in 1829, Angelina spent nine anguished months viewing slavery for the first time through eyes that judged it a sin.

Returning to Philadelphia, Angelina found no outlet for her growing antislavery passion. In 1827 a schism had split American Quakers into two groups. One, called "Hicksite" (after their leader Elias Hicks, a Long Island farmer) continued the traditional Quaker belief in the power of individual conscience over all other sources of authority, sacred or secular. The others called themselves "Orthodox" but actually were innovators who were adopting a creed that took precedence over individual conscience.[15] For example, they began to require members to profess a belief in the divinity of Jesus. In the great social issues of the day,

[14]For the endorsement of women's spiritual leadership within Quakerism, see Janis Calvo, "Quaker Women Ministers in Nineteenth Century America," *Quaker History* 63, no. 2 (1974): 75–93; and Margaret Hope Bacon, *Mothers of Feminism: The Story of Quaker Women in America* (New York: Harper, 1986).

[15]See Thomas D. Hamm, *The Transformation of American Quakerism: Orthodox Friends, 1800–1907* (Bloomington: Indiana University Press, 1988).

including slavery, Hicksite Quakers tended to promote radical reform and Orthodox Quakers tried to avoid controversy. Angelina and Sarah's friends were affiliated with the Arch Street Meeting House, seat of Orthodox Quaker conservatism. "We mingle almost entirely with a Society which appears to know but little of what is going on outside of its own immediate precincts," Angelina wrote upon her return to Philadelphia in 1829. By 1836 she put it more strongly, "My spirit is oppressed and heavy laden, and shut up in prison."[16]

Cut off from the city around her, Angelina Grimké withdrew into a five-year moratorium between late 1829 and early 1835, during which she never spoke in meeting. Outwardly manifesting no opinions about herself or her society, she drifted in a protective cloud removed from her origins and from any need to declare her views. When she emerged from this retreat she was reborn as a person who sought not to adapt herself to southern or northern opinion, but to change both regions to conform to her view of right and wrong.[17]

This isolation and the sisters' initial pursuit of religious rather than antislavery goals account for their failure in the early 1830s to ally with Lucretia Mott (1793–1880), a Hicksite Quaker minister and the leading abolitionist in Philadelphia. In 1831 Mott befriended William Lloyd Garrison, an antislavery journalist who that year founded *The Liberator,* the nation's first newspaper to call for immediate and unconditional emancipation of slaves. (See Figure 1.) Mott helped Garrison develop a more effective speaking and writing style. In 1833 she participated in the founding of Garrison's militant new organization, the American Anti-Slavery Society (AASS), and that year she also helped create one of the largest new Garrisonian women's organizations, the Philadelphia Female Anti-Slavery Society.[18] (See Document 1.)

Large numbers of women responded to Garrison's new movement even before the Grimkés joined their ranks in 1835.[19] Three white

[16]AG to Thomas Grimké (1829); and AG to SG (1836), quoted in Birney, *The Grimké Sisters,* 91 and 137.

[17]For another example of a moratorium experienced by a young person who emerged from it to change his society, see Erik Erikson, *Young Man Luther* (New York: Norton, 1958).

[18]For Garrison and *The Liberator,* see William E. Cain, ed., *William Lloyd Garrison and the Fight against Slavery: Selections from* The Liberator (Boston: Bedford Books, 1995). For Mott, see Margaret Hope Bacon, *Valiant Friend: The Life of Lucretia Mott* (New York: Walker, 1980); and Dana Greene, ed., *Lucretia Mott: Her Complete Speeches and Sermons* (New York: Mellen, 1980).

[19]For the growth of women's antislavery organizations in the 1830s, see Julie Roy Jeffrey, *The Great Silent Army of Abolitionism: Ordinary Women in the Antislavery Movement* (Chapel Hill: University of North Carolina Press, 1998), 53–95.

Figure 1 William Lloyd Garrison at age 30 in 1835, the year Angelina Grimké met him.
Courtesy of the Trustees of The Boston Public Library.

women, three black men, and sixty white men were present at the founding of the AASS, a meeting that was not publicly announced because its organizers feared the mobs that had plagued Garrison's public appearances ever since he called for the unconditional abolition of slavery. Although women could join the AASS and its local branches, women also formed separate female organizations, like the Philadelphia Female Anti-Slavery Society, which included white and black women among its founders. Such public action was new for women. In the 1820s women had begun to create maternal societies where they discussed the new meaning of motherhood in their culture, but before 1835 these groups usually met under religious auspices with ministerial leadership. Lucretia Mott, though a gifted speaker, did not feel capable of chairing the founding meeting of the Philadelphia Female Anti-Slavery Society. Yet she did not follow the common path and ask a white minister to preside; women who attended that meeting were led by "James McCrummel, a colored man." (See Document 1.)

Although black women joined the Philadelphia and other female antislavery societies, they also formed their own groups. Indeed in Massachusetts in 1832, black women formed the Female Anti-Slavery Society of Salem, the nation's first women's antislavery group. Before Garrison began to reach African Americans through *The Liberator,* many were politicized by David Walker's *Appeal to the Colored Citizens of the World,* an 1829 pamphlet that predicted and urged armed rebellion in the South. An African American merchant in Boston, Walker was a member of the Massachusetts General Colored Association, formed in 1826 to combat racial prejudice in the North and slavery in the South.

Most black women's organizations stressed racial betterment as well as the abolition of slavery. For example, the Afric-American Female Intelligence Society of America, founded in Boston in 1831, opposed slavery, but its constitution emphasized "the welfare of our friends." (See Document 2.) While many white and black women abolitionists spoke out against racial prejudice, black women dedicated themselves to the improvement of African American communities — a project that distinguished them from most, though not all, white abolitionist women. In the 1830s this emphasis within black women's organizations meant that racial justice became a higher priority for them than the advancement of women's rights. Although black women were prominent in the abolitionist movement, their voices were not generally heard on the question of women's rights until the 1850s.[20]

Nevertheless, one African American woman's campaign in Boston between 1831 and 1833 offers a fascinating exception to this rule. Maria Stewart, a black Bostonian, became the first American woman to speak in public to a mixed audience of men and women, addressing a gathering at Boston's Franklin Hall in 1832. In 1831 Stewart had taken some of her essays to Garrison at *The Liberator;* he published them and encour-

[20]See Shirley J. Yee, *Black Women Abolitionists: A Study in Activism, 1828–1860* (Knoxville: University of Tennessee Press, 1992), 136–54.

aged her to write more. Speaking before mixed audiences of men and women, white and black, Stewart urged women to unite to improve their own lives, the lives of their children, and their communities. Because the black community was more familiar with lay women preachers, her public appearances were initially tolerated. Eventually, however, criticisms drove her from Boston.[21]

Religiously motivated, Stewart preached racial and gender equality. She especially denounced the economic discrimination that segregated blacks into an exploited servant class, demanding, "How long shall the fair daughters of Africa be compelled to bury their minds and talents beneath a load of iron pots and kettles?" Stewart urged black women who maintained their own businesses to hire black girls, and she urged white women to recognize how they helped to construct racial prejudice in their daily domestic and social interactions. (See Documents 3, 4, and 5.)

For herself, Maria Stewart claimed the right to speak by exercising that right. As was later the case with the Grimké sisters, religious beliefs sustained her willingness to violate custom. Yet she never fully articulated her right to speak to groups of men and women. She urged women to speak out in defense of black rights, not women's rights. Unable to find support for her efforts in either the black or white communities, Stewart left Boston in 1833, bitterly disappointed that she had labored "in vain." (See Document 5.) She moved to New York City and rarely spoke in public again.

Maria Stewart's lectures in Boston occurred as a wave of proslavery, antiblack violence hit the urban North in response to Garrison's militant advocacy of immediate, unconditional abolition, and its accompanying call for racial equality. Heretofore white northern and southern antislavery opinion had been dominated by gradualists in the American Colonization Society who viewed blacks as inferior and had no immediate plans for their emancipation. They imagined emancipation would occur only after slave owners were compensated for the loss of their slave "property," and expected freed slaves eventually to settle in Africa. Colonization did not threaten the economic, political, and social status quo. Garrison's uncompromising call for immediate abolition did. By emphasizing the equality of blacks and whites, by attacking the racial basis

[21] For a complete account of Stewart's life, see Marilyn Richardson, ed., *Maria W. Stewart, America's First Black Woman Political Writer: Essays and Speeches* (Bloomington: University of Indiana Press, 1987). See also Carla L. Peterson, *"Doers of the Word": African-American Women Speakers and Writers in the North (1830–1880)* (New York: Oxford University Press, 1995), 56–73.

of slavery, and by denouncing race prejudice in the North, Garrison opened a new era in the nation's history.[22]

Expressing the strong economic ties that bound northern commerce to the south and the stereotypes that viewed black workers, whether slave or free, as unfair competitors with whites, mobs that included "gentlemen of property and standing" violently disrupted abolitionist meetings from Boston to Cincinnati.[23] In 1835, when the AASS sponsored a speaking tour by British abolitionist George Thompson to celebrate Parliament's recent abolition of slavery in the British West Indies, mobs greeted his appearance throughout the Northeast.

The power of Garrison's vision flowed from his contact with free blacks, who never accepted colonization as a strategy for ending slavery. The radicalism of Hicksite Quakers also shaped his views. A third force that fueled the appeal of Garrison's ideas was the Second Great Awakening, a diverse religious movement between 1800 and 1860 that transformed and revitalized Protestantism in the United States. This religious change was especially important because it affected vast numbers of Americans in the mainstream Protestant denominations, while free blacks and Hicksite Quakers were relatively few in number. The Second Great Awakening emphasized the sacred power of the individual will. Named for the "Great Awakening" of the 1730s and 1740s, which located religion in individual feelings rather than theological arguments, the Second Great Awakening overturned secular, Enlightenment attitudes of the Revolutionary era. Congregationalists (the New England denomination begun by the Puritans), Presbyterians, Baptists, and Methodists were all profoundly altered by the evangelical "good news" that individuals could will their own conversion from sin and achieve salvation. The liberating message of the Second Great Awakening released an explosion of energy in the 1830s and 1840s, as average people believed themselves able to construct better lives for themselves, their families, and their communities. In its most extreme form, this doctrine became perfectionism — belief in the human ability to become perfectly free of sin. Also called "ultraists," perfectionists often viewed secular and religious authority as corrupt and unnecessary.

The Second Great Awakening's power was fueled significantly by an emerging middle class, which consolidated its position as the society's dominant group by affiliating with the Awakening's moral authority. The power of middle-class church members was also enhanced by the sepa-

[22] On this theme, see Paul Goodman, *Of One Blood: Abolitionism and the Origins of Racial Equality* (Berkeley: University of California Press, 1998).

[23] Leonard L. Richards, *"Gentlemen of Property and Standing": Anti-Abolition Mobs in Jacksonian America* (New York: Oxford University Press, 1970).

ration of church and state that occurred throughout the new nation in the early decades of the nineteenth century. When ministers ceased to be supported by taxation and had to rely instead on voluntary contributions, many preached a gospel that exalted the voluntary efforts of their congregations, especially their women congregants.[24] These doctrinal and social changes challenged the traditional authority of ministers at the same time that they unleashed enormous energy within the new nation's civil society.

Different groups and individuals applied the lessons of the Awakening in different ways. Catharine Beecher and others who supported her brand of New England, evangelical conservatism sought to use the force of the Awakening to enhance the power of the emerging, white, middle class. They promoted a new social order, but one with distinct hierarchies of authority.

Garrison and his supporters had more egalitarian goals. A small minority in the 1830s, they argued that slavery was a sin because it deprived human beings of the freedom they needed to choose their own salvation. Because all humans were moral beings, created free by God to determine their own salvation, no person could rightfully deprive others the freedom to make moral choices. To do so was a sinful abomination. The remedy? Immediate and unconditional abolition.

Garrison's new form of antislavery activism was based on a radical religious commitment to human equality. Through the American Anti-Slavery Society and *The Liberator,* Garrison and his supporters used the language of insurgent New England Evangelicalism to sacralize slaves — that is, to make them holy. In Garrison's movement the slave became a child of God, created equal to all others in the eyes of God, and the abolition of slavery became the means by which the new nation could achieve a higher and purer form of God-given liberty.[25]

Initially Garrison's radical views about racial equality attracted far more free blacks than whites. In the 1830s free blacks rather than white evangelicals constituted the majority of the subscribers to *The Libera-*

[24]See Whitney R. Cross, *The Burned-Over District: The Social and Intellectual History of Enthusiastic Religion in Western New York, 1800–1850* (Ithaca: Cornell University Press, 1950); Nathan O. Hatch, *The Democratization of American Christianity* (New Haven: Yale University Press, 1989); Nancy A. Hardesty, *Your Daughters Shall Prophesy: Revivalism and Feminism in the Age of Finney* (Brooklyn: Carlson, 1991); and Kathryn Kish Sklar, *Catharine Beecher: A Study in American Domesticity* (New Haven: Yale University Press, 1973).

[25]See Robert H. Abzug, *Cosmos Crumbling: American Reform and the Religious Imagination* (New York: Oxford University Press, 1994), 129–62; and Caroline L. Shanks, "The Biblication Anti-Slavery Argument of the Decade, 1830–1840," *Journal of Negro History* 16 (April 1931).

tor.[26] The abolition of slavery in the North did not necessarily confer citizenship on free blacks. By 1840 more than 90 percent of northern free blacks lived in states that denied them the right to vote.[27] In his 1832 book, *Thoughts on African Colonization,* Garrison scornfully asked those who believed that blacks and whites could never live together harmoniously: "Are we pagans, are we savages, are we devils?" He insisted, "there *is* power enough in the religion of Jesus Christ to melt down the most stubborn prejudices, to overthrow the highest walls of partition, to break the strongest caste, to improve and elevate the most degraded, to unite in fellowship the most hostile, and to equalize and bless all its recipients." Quoting scripture, he declared: "In Christ Jesus, all are one: there is neither Jew nor Greek, there is neither bond nor free, there is neither male nor female."[28] This passage supported Garrison's commitment to racial equality; it also endorsed a related principle of sexual equality, which the Grimké sisters would soon explore.

Angelina Grimké was jarred out of her moratorium in 1834, when she responded to Garrison's efforts to carry a new, militant antislavery message into the South as well as the North. Despite the disapproval of her Orthodox Quaker community, she heard George Thompson, a prominent British abolitionist, speak in Philadelphia in March 1835. That year she also attended meetings of the Philadelphia Female Anti-Slavery Society, which had cosponsored Thompson's talk. She began to read *The Liberator,* where she found thrilling accounts of the heroism and martyrdom of abolitionists threatened by angry mobs. (See Figure 2.)

In the summer of 1835, in a step that unofficially severed her ties with her respectable Quaker community, Angelina Grimké publicly joined the vilified abolitionist movement. She did so just after Garrisonians carried their message to Charleston. The AASS had launched a "postal campaign" that sent vast amounts of abolitionist literature through the mails that summer—175,000 pieces through the New York City post office in July alone.[29] Mobs reacted by destroying these "inflammatory appeals" and attacking abolitionist meetings. In Charleston a mob broke into the post office, seized the AASS literature, and burned it beneath a hanged effigy of Garrison. With the bonfires in Charleston on her mind, in August Angelina sent a letter to Garrison that emphasized her profoundly

[26]Abzug, *Cosmos Crumbling,* 136.

[27]Litwack, *North of Slavery,* 75.

[28]William Lloyd Garrison, *Thoughts on African Colonization* (Boston, 1832); reprinted in George M. Fredrickson, ed., *William Lloyd Garrison* (Englewood Cliffs, N.J.: Prentice-Hall, 1968), 29–37, quote 33.

[29]Bertram Wyatt-Brown, "The Abolitionists' Postal Campaign of 1835," *Journal of Negro History* 50 (Oct. 1963): 227–38.

Figure 2 Angelina Grimké at about age 39, c. 1845. Her piercing eyes and firm countenance conveyed conviction to those in her 1830s audiences.
Boston Athenæum.

personal commitment, saying, "it is my deep, solemn, deliberate conviction, that *this is a cause worth dying for.*" She explained,

> O,! how earnestly have I desired, *not* that we may escape suffering, but that we may be willing to endure unto the end. If we call upon the slaveholder to suffer the loss of what he calls property, then let us show him we make this demand from a deep sense of duty, by being ourselves willing to suffer the loss of character, property—yes, and life itself, in what we believe to be the cause of bleeding humanity.[30]

[30] AG to William Lloyd Garrison, Aug. 30, 1835, printed in *The Liberator,* Sept. 19, 1835, and reprinted in Ceplair, *The Public Years,* 24–27, quote 26. For AG's state of mind, see AG to SG, Sept. 27, 1835, quoted in Birney, *The Grimké Sisters,* 127.

Angelina Grimké had found her voice. She was thirty years old. At that moment Angelina joined Garrison, Mott, and others in the vanguard of her generation's concern for the nation's future as a land of liberty.

Three Americans who later played major parts in the national drama over slavery were relatively oblivious to its importance in 1835. Abraham Lincoln, twenty-six years old, was a freshman state legislator and law student in New Salem, Illinois. Jefferson Davis, twenty-six years old and later President of the Confederate States of America, had just moved with his bride and eleven slaves to begin a cotton plantation on rich Mississippi delta land. Harriet Beecher Stowe, twenty-four years old and future author of *Uncle Tom's Cabin* in 1852, was teaching school in Cincinnati. When Lincoln, Davis, and Stowe later became large public figures, they joined a discourse that had been shaped by the Grimké sisters and other Garrisonians.

WOMEN CLAIM THE RIGHT TO ACT: ANGELINA AND SARAH GRIMKÉ SPEAK IN NEW YORK, JULY 1836–MAY 1837

At first, Sarah strongly disapproved of her sister's actions, writing in her diary, "The suffering which my precious sister has brought upon herself by her connection with the antislavery cause, which has been a sorrow of heart to me, is another proof how dangerous it is to slight the clear convictions of truth." Truth, for Sarah, lay in the Quaker admonition to be still and avoid conflict. (See Figure 3.) Angelina wrote her, "I feel as though my character had sustained a deep injury in the opinion of those I love and value most — how justly, they will best know at a future day."[31]

Angelina moved ahead, independently of her sister. She took refuge with a sympathetic friend in Shrewsbury, New Jersey, and spent the winter of 1836 writing *Appeal to the Christian Women of the South.* Writing as a southerner and a woman, she created a place for herself within the new Garrisonian movement. Promptly published and widely distributed by the AASS, her *Appeal* urged southern women to follow the example of northern women and mobilize against slavery. Rather than limit their influence to their domestic circles, Angelina argued, women should speak out and take public action. "Where *woman's* heart is bleeding," she insisted, "Shall *woman's* heart be hushed?" (See Figure 4.) If championing the cause of the slave required them to break local laws, she called

[31] Diary of SG, Sept. 25, 1835; and AG to SG, Sept. 27, 1835, both quoted in Birney, *The Grimké Sisters,* 128–29.

Figure 3 Sarah Grimké at about age 50, c. 1842. The sisters' close resemblance to one another reinforced the power of their appeal.
Boston Athenæum.

on her readers, in one of the nation's first expressions of civil disobedience, to follow higher laws.[32] (See Document 7.)

The exiled South Carolinian especially urged her readers to petition the national government to end slavery. The recent abolition of slavery in the British West Indies, she argued, was due to women's petitions. Sixty female antislavery societies in the North had already followed the British women's example, and she urged southern women to do the same.

[32] For artwork associated with this appeal to women, see Jean Fagan Yellin, ***Women and Sisters: The Antislavery Feminists in American Culture*** (New Haven: Yale University Press, 1989).

Figure 4 "Am I Not a Woman and a Sister?" First published on the cover of Lydia Maria Child's *Authentic Anecdotes of American Slavery* (1838), this image was one of many popular depictions of a caring but hierarchical relationship between white women abolitionists and women slaves.

By endorsing the petition movement Angelina Grimké affiliated with the most political outcome of women's abolitionist activism before 1837. Petitions enlarged the public space that women occupied, and brought them into direct contact with their national government. At a time when the national government was a distant and somewhat ephemeral concept in the lives of most Americans, antislavery petitions made it a concrete reality. Women's antislavery petitions affirmed the potential power of the national government to redress grievances and restore "rights unjustly wrested from the innocent and defenseless." (See Document 6.)

Disfranchised individuals, like women and slaves, had petitioned the U.S. Congress since the 1790s, drawing on the First Amendment to the Constitution, which guaranteed the right of "the people . . . to petition the government for a redress of grievances." What was new in the 1830s was the increase in group petitions with many signatures. The first of these was a sabbatarian campaign led by Lyman Beecher to stop the movement of U.S. mail on Sunday. A distinctly women's campaign emerged in the early 1830s, when northern women, including Catharine Beecher, petitioned against President Andrew Jackson's forced removal of Cherokee people from their ancestral lands in Georgia.[33] The AASS's

[33]Richard John, "Taking Sabbatarianism Seriously: The Postal System, the Sabbath, and the Transformation of American Political Culture," *Journal of the Early Republic* 10 (1990): 517–67; Mary Hershberger, "Mobilizing Women, Anticipating Abolition: The Struggle against Indian Removal in the 1830s," *Journal of American History* 86, no. 1 (June 1999): 15–40; Anna Goldstein, "Women's Petitions against Cherokee Removal in the 1830s," on "Women and Social Movements in the United States, 1830–1930," a Web site at http://womhist.binghamton.edu.

campaign to end slavery in the District of Columbia carried this new form of political expression into the most explosive issue in American politics.

Recognizing the growing numbers of women's antislavery societies, and the dedication of their members, the AASS printed a petition form especially designed for women. (See Document 6.) The new women's societies dedicated themselves to gathering signatures, walking door to door, and driving wagons or carriages through their communities. The first such petition, signed by "eight hundred ladies," was presented by a New York Congressman to the U.S. House of Representatives in February 1835.[34] Historians estimate that women contributed seventy percent of the signatures on antislavery petitions.[35] They collected three times as many signatures as those previously obtained by paid male AASS agents. By 1836 the petition campaign so disrupted the proceedings of the U.S. House of Representatives that a "gag rule" was passed to prevent congressmen from reading or otherwise presenting the petitions on the floor of the House. Passed with each Congress between 1836 and 1844, gag rules expanded support for the antislavery cause by linking it to free speech.

Joining this insurgent movement in 1835, Angelina Grimké immediately attracted the attention of the AASS, which sent piles of her *Appeal* to Charleston. There, postmasters publicly burned the pamphlets. The city's mayor told Mrs. Grimké that her daughter would be arrested if she tried to enter the city; if she visited, Angelina's friends wrote her, "she could not escape personal violence at the hands of the mob."[36] Justifying her actions to Sarah, Angelina acknowledged that her pamphlet was "a pretty bold step . . . of which my friends will highly disapprove, but this is a day in which I feel I must act independently of consequences to myself." The South must be reached, she felt, and "an address to men will not reach women, but an address to women will reach the whole community."[37]

[34] W. P. and F. J. Garrison, *William Lloyd Garrison, 1805–1879: The Story of His Life Told by His Children, 1: 1805–1835* (New York: Century, 1885), 482.

[35] For the petition campaign, see Gerda Lerner, "The Political Activities of Antislavery Women," in Lerner, *The Majority Finds Its Past: Placing Women in History* (New York: Oxford University Press, 1979); and Deborah Bingham Van Broekhoven, "'Let Your Names Be Enrolled': Method and Ideology in Women's Antislavery Petitioning," in *The Abolitionist Sisterhood: Women's Political Culture in Antebellum America,* ed. Jean Fagan Yellin and John C. Van Horne (Ithaca: Cornell University Press, 1994).

[36] Theodore Weld's later account, quoted in Birney, *The Grimké Sisters,* 149.

[37] AG to SG (Shrewsbury, summer 1836), quoted in Birney, *The Grimké Sisters,* 141. See also Stanley Harrold, *The Abolitionists and the South, 1831–1861* (Lexington: University Press of Kentucky, 1995).

Even before Angelina submitted her *Appeal* to the American Anti-Slavery Society, Elizur Wright, secretary of the society, invited her to come to New York and, under the sponsorship of the AASS, meet with women in their homes and speak with them about slavery. Preceded by the reputation of her brother, Thomas, whom Garrison and other AASS members knew as a member of the American Peace Society, as well as by the eminence of her deceased father, Angelina was enthusiastically welcomed into the AASS fold.

Sarah, meanwhile, was rebuked in the Arch Street Meeting House in a way that led her to leave Philadelphia. Although her speaking skills remained unimpressive, and her southern accent reminded her listeners that she remained an outsider in this elite Quaker community, Sarah often spoke in meeting. In July 1835, a presiding elder, probably expressing a consensus reached with other local leaders, rose and cut her off, saying: "I hope the Friend will now be satisfied." Silenced, Sarah sat down. His breach of Quaker discipline was clearly meant to silence her permanently in the meeting. For nine years she had struggled to develop ministerial speaking gifts, keenly aware of the cold indifference with which the elders viewed her efforts. That day she decided "that my dear Saviour designs to bring me out of this place." On learning the news, Angelina rejoiced, "I will break your bonds and set you free."[38] Within a few weeks Angelina had convinced Sarah of the righteousness of her Garrisonian views, and Sarah acknowledged the younger sister's leadership in setting their future course.

Renouncing the respectable comforts of northern as well as southern society, the sisters traveled to New York City in the autumn of 1836. There for most of the month of November 1836, they joined a training workshop of about thirty men who were serving as the paid agents of the AASS. They forged new identities as antislavery agitators. "We sit," Angelina wrote to a friend, "from 9 to 1, 3 to 5, and 7 to 9, and never feel weary at all," discussing biblical arguments against slavery and answering questions like "What is slavery?"[39]

Although they declined to accept pay from the AASS, the sisters launched their new identities as the first women agents of the AASS on December 16, 1836, when they spoke in a Baptist meeting room, no home being large enough to hold the three hundred women who wanted to hear them. To speak in public they had to be willing to break the cus-

[38]Diary of Sarah Grimké, Aug. 3, 1836; and AG to SG, both quoted in Birney, *The Grimké Sisters*, 144–45.

[39]AG to Jane Smith, Nov. 11, 1836; and AG to Jane Smith, Nov. 1836, both quoted in Birney, *The Grimké Sisters*, 159.

tom that prohibited respectable women from addressing large public gatherings, and oppose the scriptural authority of Paul's admonition to the early Christians, "Let your women keep silence in the churches; for it is not permitted unto them to speak."[40] Antislavery women had heretofore spoken to small groups of women, mostly neighbors, in their homes or churches. Angelina temporarily lost her nerve and feared it would be "unnatural" for her to proceed. Charismatic Theodore Weld, one of the leaders of their training group, whom Angelina would later marry, revived the sisters' courage by disparaging social norms that "bound up the energies of woman," and by reminding them of the high importance of their message. (See Document 8 and Figure 5.)

Public speaking was a form of performance in nineteenth-century culture that was strongly associated with the explicitly masculine virtues of virility, forcefulness, and endurance. Much of the vitality of the new nation's public life emanated from styles of oratory that began to emerge in the 1830s, first in the pulpit, then in politics. These new styles were part of the process by which the authority of the clergy and of traditional landed elites was being replaced in public life by the power of more popular groups, especially the broad middle class that ranged from artisans to professionals. The rise in public oratory placed great demands on both speaker and audience. Speakers were expected to engage their audiences' emotions in ways that entertained as well as enlightened, and they were not expected to be brief. Theirs was a culture in which "the word" carried great weight.[41]

Beyond the meetings of all-female societies, respectable women were permitted to enter this arena only as consumers, not as the producers of eloquence. The chief exceptions to this rule were itinerant women preachers, whose prophetic or visionary preaching was inspired by the new currents of expression within the Second Great Awakening. In 1820 Deborah Pierce had defended them in a pamphlet, *A Scriptural Vindiction of Female Preaching, Prophesying, or Exhortation*. Yet the authority that these women claimed was spiritual, not social. One historian has called them "Biblical Feminists" because they asserted the spiritual but not the social equality of women. Women might have mystical powers but this did not translate into authority in society. Female preachers were not ordained and worked outside the institutional religious author-

[40] I Corinthians 14:34.

[41] Although literacy was universal in the North, the spoken word still had the authority that came in part from its identification with religious authority, as in "In the beginning was the Word, and the Word was with God, and the Word was God" (John 1:1).

Figure 5 Theodore Dwight Weld at about age 38, c. 1841. Forceful and defiant, Weld was a riveting antislavery orator.
The Library Company of Philadelphia.

ity of the clergy.[42] In 1836 and 1837 Angelina and Sarah Grimké asserted women's right to speak not only because women were the spiritual equals of men, but also because they were the moral and social equals of men. By so doing, they opened new channels to public life for women who were not visionaries or prophets.

Opponents tried to discredit women's public speaking by tainting it with the radicalism of Frances Wright, a British free thinker identified

[42]See Catherine Anne Brekus, *Strangers and Pilgrims: Female Preaching in America, 1740–1845* (Chapel Hill: University of North Carolina Press, 1998).

with the French Revolution, who spoke widely in the United States in the 1820s and created a short-lived interracial utopian community in Tennessee.[43] (See Document 8.) Even the term used to describe a mixed audience of men and women, *promiscuous,* conveyed the era's distrust of women moving freely in public space.

Partly because public speaking was central to the construction of contemporary civil society, the prohibition against respectable women's public speaking was a key ingredient in the practices that denied respectable women — elite, middle-class, and artisan-class women — access to and equality within the new contours of public life. By speaking in public the Grimké sisters precipitated a women's movement that rebelled against this exclusion. By breaking the prohibition against women's public speaking, they generated criticisms that denied their right to speak, which in turn prompted them to defend that right. This spiraling dynamic occurred because thousands of people wanted to hear the sisters speak more than they wanted to enforce customary strictures.[44]

Addressing a new audience each week, Angelina's confidence in her public performances grew in the meetings that followed her first speaking date. In January she wrote her Philadelphia friend, Jane Smith, "I love the work." (See Document 9.) By February she described her animated speaking style as yielding to "impulses of feeling." By then she was also beginning to defend women's "rights & duties" to speak and act on the topic of slavery. (See Document 10.)

In addition to propelling women into public activism and greater awareness of the positive uses of government, the Grimkés' speaking tour of 1836 brought interaction with large numbers of African Americans, and gave them the opportunity to act on their Garrisonian principles against racial prejudice. "The more I mingle with your people, the more I feel for their oppressions & desire to sympathize in their sorrows," Angelina wrote Sarah Douglass, a black friend from the Arch Street Quaker Meeting.[45] (See Document 11.) In Poughkeepsie, New York, in March 1837, they spoke to a black group that included men as

[43] See A. J. G. Perkins and Theresa Wolfson, *Frances Wright, Free Enquirer: The Study of a Temperament* (New York: Harper & Bros., 1939).

[44] For more on the Grimkés' oratory, see Lillian O'Connor, *Pioneer Women Orators: Rhetoric in the Ante-Bellum Reform Movement* (New York: Vantage Press, 1952); and Karlyn Kohrs Campbell, *Man Cannot Speak for Her, 1: A Critical Study of Early Feminist Rhetoric* (New York: Praeger, 1989). For an exploration of other ways that women were present in public life, see Mary P. Ryan, *Women in Public: Between Banners and Ballots, 1825–1880* (Baltimore: Johns Hopkins University Press, 1990).

[45] For Sarah Douglass, see Marie J. Lindhorst, "Sarah Mapps Douglass: The Emergence of an African American Educator/Activist in Nineteenth Century Philadelphia," Ph.D. diss. (University Park: Pennsylvania State University, 1995).

well as women. "For the first time in my life I spoke in a promiscuous assembly," Angelina said. (See Document 12.)

Their Philadelphia Quaker community had introduced the Grimkés to northern racism at the same time that it acquainted them with some of the city's black elite and middle class. Regardless of class, African Americans who belonged to the Arch Street Meeting sat in a segregated pew. Angelina wrote Sarah Douglass, daughter of a black artisan family, that such customs now made her "feel ashamed for my country, ashamed for the church," though she thought "the time is coming when such 'respect of persons' will no more be known in our land." Angelina also asked Sarah Forten, daughter of an elite African American Philadelphia abolitionist, and a member of the Philadelphia Female Anti-Slavery Society, to educate her about the effects of racial prejudice in her life. Forten eloquently described the "weight of this evil."[46] (See Document 13.) Angelina had hoped to found a national organization of antislavery women, but gave up when she realized that racial prejudice among New York abolitionist women would banish "our colored sisters from an equal & full participation in its deliberations & labors."[47] Thus the Grimké sisters' perception of the human rights of their "colored sisters" evolved alongside their perception of "the rights and duties of women."

On this topic the South Carolina exiles tried to improve their white sisters by meeting with the board of the New York Female Anti-Slavery Society. "I believed it right to throw before them our views on the state of things among them, particularly on prejudice," Angelina wrote Jane Smith. "No colored Sister has ever been in the board, & they have hardly any colored members." The meeting was emotional. "What we said to them was from a sense of duty in love & tears, but it was hard work, & I believe as much as they could possibly hear from us. But some were reached, I do believe, for tears were shed."[48] In this spirit in 1837 Sarah wrote *Address to the Free Colored People of the United States,* which declared, "In the sight of God, and in our own estimation, we have no superiority over you."[49]

The rapid growth of women's antislavery associations, and the stir that the Grimkés were making in its ranks, led to an unprecedented

[46] Sarah Forten's niece, Charlotte Forten, kept a diary of her life in Philadelphia and among freedmen in the South, published as Ray Allen Billington, ed., *The Journal of Charlotte L. Forten* (New York: Macmillan, 1961). In 1878 Charlotte married Francis Grimké, the son of AG and SG's brother Henry Grimké, and a slave, Nancy Weston.

[47] AG to Jane Smith, New York, March 22, 1837, quoted in Ceplair, *The Public Years,* 126.

[48] AG to Jane Smith, March 22, 1837.

[49] Sarah M. Grimké, *Address to the Free Colored People of the United States* (Philadelphia: Merrihew and Gunn, 1837), 3.

Figure 6 Maria Weston Chapman at about age 41, c. 1847. Chapman was Garrison's chief lieutenant in Massachusetts, who helped him edit *The Liberator.* She vigorously supported the Grimké sisters.
Boston Athenæum.

event in May 1837—a national convention of antislavery women. Organized by Angelina Grimké and other women abolitionist leaders, including Maria Weston Chapman of the Boston Female Anti-Slavery Society (see Figure 6) and Lucretia Mott and others in the Philadelphia Female Anti-Slavery Society, the convention was scheduled in conjunction with the annual meeting of the American Anti-Slavery Society. Planning for the meeting, Angelina urged Sarah Douglass to attend it despite the prejudice she could expect to encounter there—from women abolitionists as well as the public at large. (See Document 12.) Accompanied by her mother, Sarah Douglass did attend. The three-day event attracted about 200 women from nine states. About one in ten was African American.

The women's antislavery convention marked a new stage in the emergence of women's rights within the abolitionist movement. There Angelina presented her *Appeal to the Women of the Nominally Free States,* a lengthy pamphlet that defiantly defended women's right to speak and act. "The denial of our duty to act, is a bold denial of our right to act; and if we have no right to act, then may *we* well be termed 'the white slaves of the North'—for, like our brethren in bonds, we must seal our lips in silence and despair." (See Document 14.) Grimké's *Appeal* also contained a Garrisonian assault on race prejudice.

At the convention women adopted resolutions that urged women to circulate petitions annually, that decried the indifference of U.S. churches to the sin of slavery, that censured northern women and men who married southern slaveholders, and that called for white women to associate with African American women "as though the color of the skin was of no more consequence than that of the hair, or the eyes."[50] Just as radically, the convention approved a resolution supporting women's rights. It declared that "certain rights and duties are common to all moral beings," and that it was the duty of woman "to do all that she can by her voice, and her pen, and her purse, and the influence of her example, to overthrow the horrible system of American slavery." Although most of the convention's resolutions were adopted without extensive discussion, this one "called forth an animated and interesting debate." Twelve delegates opposed the resolution so strongly that they had their names recorded in the minutes "as disapproving." (See Document 15.) Not all women abolitionists were willing to be associated with the radical claim that women and men had equal rights. Many thought that their mere attendance at the convention was sufficiently daring. "To attend a Female convention!" one exclaimed. "Once I should have blushed at the thought."[51]

Angelina thought that the debate at the convention was healthy, since it "very soon broke down all stiffness & reserve, threw open our hearts

[50]The full proceedings of the convention have been published as *Turning the World Upside Down: The Anti-Slavery Convention of American Women, Held in New York City, May 9–12, 1837* (New York: Feminist Press, 1987).

[51]Quoted in Dorothy Sterling, *Ahead of Her Time: Abby Kelley and the Politics of Antislavery* (New York: Norton, 1991), 44.

to each others' view, and produced a degree of confidence in ourselves & each other which was very essential & delightful." She predicted that their resolutions would "frighten the weak and startle the slumbering, particularly those on Southern intermarriages, the province of women, the right of Petition."[52] Members of the Philadelphia Female Anti-Slavery Society credited the convention for their increased membership, their robust treasury, and their renewed petition campaign. (See Document 30.)

Although two other conventions of antislavery women were subsequently held in Philadelphia, in 1838 and in 1839, no resolutions on women's rights and duties were offered there, perhaps because they were perceived as too divisive. Nevertheless, by 1837 women's rights ideas had taken root within the antislavery movement. Critics of women's public activism kept the topic in the public spotlight and escalated the scope and intensity of the debate.

The first of these critics was Catharine Beecher, whose 1837 book, *An Essay on Slavery and Abolitionism, with Reference to the Duty of American Females,* appeared just before the May convention as a reply to Angelina Grimké's *Appeal to the Christian Women of the South.* (See Document 16.) Beecher had moved with her father and siblings to the Ohio frontier settlement of Cincinnati, seeking to claim the West for New England Evangelicalism. Like most of her family at that time, she viewed Garrisonian abolitionism as a disruptive force in American society that imperiled their efforts to reconstitute the Northwest Territory along New England cultural lines. She was not so much a defender of traditional male privilege as a competitor with the Grimkés for the loyalty of middle-class evangelical women. Her *Essay* argued that the abolitionist emphasis on immediate emancipation was shutting off communication between the North and South, and might lead to the creation of a "Southern republic" with slaveholding preserved in perpetuity. Seeking to build women's power in family life, she opposed women's participation in the abolitionist movement because "the attitude of a combatant" threw women out of their appropriate sphere — the "domestic and social circle." Beecher seized this opportunity to make her own goals — particularly the feminization of the teaching profession — appear conservative and respectable in contrast to abolitionist women.

Both Beecher and the Grimkés were trying to construct female identities that would permit women to assume new public responsibilities within the larger male-dominated social projects with which they

[52] AG to Jane Smith, New York, May 20, 1837, in Ceplair, *The Public Years,* 133.

were affiliated—Beecher with evangelical Congregationalism and the Grimkés with Garrisonian abolitionism. These two competing agendas within the Second Great Awakening had first clashed in 1834, at Lane Theological Seminary in Cincinnati, where Lyman Beecher was president and Theodore Weld and other abolitionist radicals were students. After Beecher told students not to mingle socially with Cincinnati's free blacks, Weld and his supporters denounced him and left the city, and joined the newly founded Oberlin College.[53] Three years later, as the Grimké sisters headed into their historic tour of New England, they knew they had to defend women's rights in terms that refuted Catharine Beecher's views.

REDEFINING THE RIGHTS OF WOMEN: ANGELINA AND SARAH GRIMKÉ SPEAK IN MASSACHUSETTS, SUMMER 1837

Following the inaugural meeting of the Anti-Slavery Convention of American Women, the Grimkés launched their historic speaking tour of Massachusetts in the summer of 1837. Beginning in Boston, they felt that they pleaded "not the cause of the slave only" but "the cause of woman as a responsible moral being." In parlor meetings where they strategized with their supporters, men as well as women thought it was time their "fetters were broken." Many believed "that women could not perform their duties as moral beings, under the existing state of public sentiment." And very many thought "that a new order of things is very desirable in this respect." Angelina was stunned by this turn of events. "What an untrodden path we have entered upon!" she wrote Jane Smith. "Sometimes I feel almost bewildered, amazed, confounded & wonder by what strange concatenation of events I came to be where I am & what I am." (See Document 17.)

Reinforced by this support, Angelina declared before 300 members of the Female Moral Reform Society in Boston that they should see women's rights as a personal issue, "that this reform was to begin in *ourselves.*" Women were "polluted" by the attitudes toward them, she said. For example, they felt "restraint & embarrassment" in the society of men. And the "solemn & sacred subject of marriage" was discussed in unseemly ways. Emphasizing the subjective, personal impact of gender inequalities, Angelina told her audience, "My heart is pained, my wom-

[53] Robert Abzug, *Passionate Liberator: Theodore Dwight Weld and the Dilemma of Reform* (New York: Oxford University Press, 1980), 74–121.

anhood is insulted, my moral being is outraged continually." (See Document 17.)

The sisters found support for this critique of male dominance when they left Boston to lecture in the surrounding countryside. Maria Weston Chapman, an elite leader of the Boston Female Anti-Slavery Society, sent a letter "to Female Anti-Slavery Societies throughout New England" to support the Grimkés' discussion of "the condition of woman; her duties and her consequent rights." Disdaining those "who were grinding in the narrow mill of a corrupt publick opinion on this point," Chapman urged antislavery women to show the sisters "hospitality of the heart."[54] (See Document 18.)

People came to hear the sisters because they were curious about the phenomenon they now represented: southern women speaking in public about their firsthand view of the horrors of slavery and about their right to speak in public about those horrors. Their reception varied. "Great apathy" sometimes reigned in small outlying towns, but before large audiences in Boston and Lynn, Massachusetts, Angelina found it "very easy to speak because there was great openness to hear." Men became an increasing part of their audiences. In Lynn on the evening of June 21 they addressed 500 women; a month later in that manufacturing town they spoke to 1,500 men as well as women. (See Documents 19 and 24.) Seven months after their first speaking engagement they had successfully broken the customs against women speaking in public and against women speaking to mixed audiences.

The Grimkés maintained a grueling pace between June and November. Speaking every other day, sometimes twice a day, often for two hours at a time, they reached thousands. Angelina's record in her letters to Jane Smith shows that in June she spoke seventeen times in ten towns, to more than eight thousand people. In July she gave nineteen lectures in fourteen towns, reaching nearly twelve thousand. In August, even though Angelina was ill for half the month, she gave eleven lectures in nine towns with six thousand present. In September, she spoke seventeen times in sixteen towns with more than seven thousand persons attending. In October she spoke fifteen times in fifteen towns, with twelve thousand in her audiences. During these five months she lectured seventy-nine times to audiences that totaled more than forty thousand people.[55]

[54] On Chapman and the Boston society, see Debra Gold Hansen, *Strained Sisterhood: Gender and Class in the Boston Female Anti-Slavery Society* (Amherst: University of Massachusetts Press, 1993); and Blanche Glassman Hersh, *The Slavery of Sex: Feminist-Abolitionists in America* (Urbana: University of Illinois Press, 1978).

[55] Lumkin, *The Emancipation of Angelina Grimké*, 128.

Since Sarah kept no equivalent record, we do not know if she ever spoke independently of Angelina, but it seems she did not. Angelina was the one people came to hear. Her extraordinary oratory offered audiences the 1830s equivalent of an award-winning movie about slavery. "Never before or since have I seen an audience so held and so moved by any public speaker, man or woman," said a Massachusetts minister in whose pulpit she lectured.[56] Wendell Phillips, a prominent Boston abolitionist, said that she "swept the cords of the human heart with a power that has never been surpassed, and rarely equalled." Phillips was impressed by "her serene indifference to the judgment of those about her. Self-poised, she seemed morally sufficient to herself." He thought her power derived from "the profound religious experience of one who had broken out of the charmed circle, and whose intense earnestness melted all opposition." Audiences felt that "she was opening some secret record of her own experience"; their "painful silence and breathless interest told the deep effect and lasting impression her words were making."[57]

Angelina's impact on her audiences issued partly from her mastery of an oratorical style that emphasized her feelings. Equally important, however, was her ability to use familiar language in new ways. She spoke in the religious discourse of her time, changing it to fit her purposes. "Cast out first the spirit of slavery from your own hearts," she said. "The great men of this country" and the "church" have become "worldly-wise, and therefore God, in his wisdom, employs them not to carry on his plans of reformation and salvation." Instead, he has chosen "the weak to overcome the mighty." This use of familiar metaphors to convey new ideas had deep roots in Judeo-Christian traditions. Angelina stood in a long line of prophets who used that tradition to bring new concepts into their cultures. Like them, she was denounced by established religious authorities, who saw her as a defiant challenge to their leadership. She equated herself to Isaiah and the Massachusetts clergy to Old-Testament Jewish priests. (See Document 28.)

Yet despite Angelina's greater oratorical gifts, the sisters' success was achieved jointly; neither could have done alone what they were able to accomplish together. Traveling among strangers, some of whom were friendly, some hostile, they could rely on one another as trusted family members and as dedicated colleagues with whom they shared the

[56] Birney, *The Grimké Sisters,* 190.

[57] [Theodore D. Weld], *In Memory of Angelina Grimké Weld* (Boston: George Ellis, 1880). For an example of AG's speaking style, see Document 37, the only one of her lectures to be recorded by shorthand.

speaking platform. Neither had to bear all the burdens of their grueling schedule. Sarah was becoming a capable speaker on women's rights, and Angelina relied on her to elaborate that aspect of their message. "Sister Sarah does preach up woman's rights most nobly & fearlessly," Angelina wrote in July. When Sarah's voice failed due to a cold, Angelina complained that she "had to bear the brunt of the meeting" for two days. (See Document 20.)

One measure of the sisters' success was the exponential growth in membership of the American Anti-Slavery Society immediately after their speaking tour. Each meeting harvested new "subscribers" to the AASS. In 1837 the AASS claimed 1,000 auxiliaries and 100,000 members. A year later their membership had more than doubled to 250,000, and the number of affiliated antislavery societies had increased by a third.[58]

A large number of these new members were women. Many women responded enthusiastically to the Grimkés' "breach in the wall of public opinion." Angelina wrote Jane Smith, "we find that many of our New England sisters are ready to receive these strange doctrines, feeling as they do, that our whole sex needs an emancipation from the thraldom of public opinion." In villages as well as cities, "the whole land seem[ed] roused to discussion on the *province of woman.*" (See Document 20.) While the sisters from South Carolina led the way, others willingly followed. Disciples attracted to their side—like Abby Kelley (1810–87), a Quaker teacher in Lynn, Massachusetts—now embraced women's rights as well as Garrisonian abolition.

In June the Grimkés' campaign created a constituency for women's rights; in July their opponents emerged. Men within the Garrisonian movement, many of whom supported women's rights in the abstract, feared the issue would damage abolitionism by diffusing the movement's energies, and by linking antislavery with an even less popular idea. Sarah responded to this criticism in a letter to Amos Phelps, a Congregational minister who was an agent of the AASS. There was no going back on her right to speak to mixed assemblies, she said. "To close the doors *now* against our brethren wd. be a violation of our fundamental principle that man & woman are created equal, & have the same duties & the same responsibilities as moral beings. If, therefore, it is right for thee, my dear brother, to lecture to promiscuous assemblies, it is right for us to do the same." (See Document 21.)

[58] Gilbert Hobbs Barnes, *The Antislavery Impulse, 1830–1844* (New York: Appleton-Century, 1933; reprint, Smith, 1957), 134–35; and Louis Filler, *The Crusade against Slavery, 1830–1860* (New York: Harper, 1960), 67.

The sharpest criticism of the sisters came from a Pastoral Letter issued by the General Association of Massachusetts (Congregational) Churches, which in July condemned "those who encourage females to bear an obtrusive and ostentatious part in measures of reform, and countenance any of that sex who so far forget themselves as to itinerate in the character of public lecturers and teachers." The letter especially condemned the naming of "things which ought not to be named," as they delicately referred to the sisters' testimonials about the sexual exploitation of slave women. Indirectly the letter forbade clergymen to permit the sisters to speak in their churches. (See Document 22.)

In contrast to most British churchmen, who led public opinion in supporting abolition in the 1830s when Parliament ended slavery in the British West Indies, most U.S. clergymen did not support abolition before the U.S. Civil War. Even after most Protestant denominations split into southern and northern branches (Presbyterians in 1837), northern clerical opinion expressed stronger opposition to Garrisonian radicalism than to slavery.[59] Exceptions to this rule included ministers who invited the sisters to speak in their churches, as well as clergymen like Henry Clarke Wright who were AASS agents.

Clerical opposition to the Grimkés was fueled in part by ministers' desire to contain the power of the female laity in their own congregations. Historically excluded from leadership positions, except among Quakers, in the late 1830s women began to express views on a wide range of issues in their communities, including temperance and moral reform as well as abolition. The temperance movement strove to limit the consumption of alcohol. The moral reform movement sought to control men's sexual behavior; it attacked the sexual double standard that condoned sexual behavior by men that was not allowed by women.[60] The growing autonomy of women's voices on these and other issues threatened to undercut ministers' moral leadership. The Reverend Albert Folsom of Hingham, Massachusetts, spoke for many clergymen when he

[59]See John R. McKivigan, *The War against Proslavery Religion: Abolitionism and the Northern Churches, 1830–1865* (Ithaca: Cornell University Press, 1984).

[60]For the temperance movement, see Ruth M. Alexander, "'We Are Engaged as a Band of Sisters': Class and Domesticity in the Washingtonian Temperance Movement, 1840–1850," *Journal of American History* 75 (Dec. 1988), 763–85. For moral reform, see Carroll Smith Rosenberg, "Beauty, the Beast, and the Militant Woman: A Case Study in Sex Roles and Social Stress in Jacksonian America," in *Women and Power in American History: A Reader: Vol. I to 1880,* ed. Kathryn Kish Sklar and Thomas Dublin (Englewood Cliffs, N.J.: Prentice-Hall, 1991), 185–98; and Daniel Wright, "What Was the Appeal of Moral Reform to Antebellum Northern Women?" in "Women and Social Movements in the United States, 1830–1930," a website at http://womhist.binghamton.edu.

sought to discredit women's participation in public life as "inappropriate and unlawful." (See Document 23.)

By August, Sarah felt that "a storm was gathering all around against our *womanhood*," and feared that even more ministers would close church doors against them. (See Document 24.) She and Angelina tried to identify their allies among antislavery men. Their strongest supporter was Henry Clarke Wright, who had trained with them in New York in November 1836. When they arrived in Boston in June 1837, the sisters had stayed in the Wright home, where, Angelina said, "we enjoyed the green pastures of christian intercourse & the still waters of *peace* in his lovely family." Angelina praised him to Jane Smith as "one of the best men I ever met with" and "one of the holyest men I ever saw."[61] Wright arranged most of the details of the sisters' speaking tour that summer, and in *The Liberator* he defended the sisters' discussion of women's rights.

Because Wright was in close daily contact with the Grimkés and because he supported other reforms in addition to antislavery, Theodore Weld and others blamed him for the sisters' diversion into women's rights. Although no one else knew it, not even Angelina, Weld was by this time in love with her, and jealously resented Wright's influence with Angelina. At the end of July, probably at Weld's insistence, the AASS Executive Committee transferred Wright to Pennsylvania. "I would gladly stay & avert every arrow of scorn & obloquy from their devoted heads," Wright noted in his diary, but the sisters had to proceed without him, even though his replacement did not approve of them and could not therefore be relied upon to book their meetings.[62] "*This burden we did not expect to have to bear, and it is often perplexing*," Angelina emphatically complained to Weld.[63]

In ways that might have expressed her desire to know Weld more intimately, Angelina's letters to him in August invited him to answer questions about his views of equality in marriage. The first of these pleaded with him to join her struggle against the opponents of women's rights. (See Document 25.) Angelina acknowledged the personal, subjective nature of women's rights, since it touched "every man's interests at

[61]AG to Jane Smith, June 26, July 16, and July 25, 1837, quoted in Ceplair, *The Public Years*, 140.

[62]Wright's diary, quoted in Ceplair, *The Public Years*, 140.

[63]AG to Theodore Weld, Fitchburg (Mass.), Sept. 20, 1837, in *Letters of Theodore Dwight Weld, Angelina Grimké Weld and Sarah Grimké, 1822–1844*, ed. Gilbert H. and Dwight L. Dumond (New York: Appleton-Century-Crofts, 1934; reprint, Gloucester, Mass.: Smith, 1965), 1: 451. [Hereafter *Weld-Grimké Letters*.]

home, in the tenderest relation of life." She and Sarah had not sought this controversy, she emphasized. "We are placed very unexpectedly in a very trying situation, in the forefront of an entirely new contest—a contest for the *rights of woman* as a moral, intelligent & responsible being." Crossing her letter in the mail, one from him must have greatly disappointed her because it urged her to give up the women's rights campaign. (See Documents 25 and 26.)

While Angelina was writing a response to Catharine Beecher's *Essay* in July and August, Sarah began to publish a series of essays in defense of women's rights. Weld urged them both to cease such publications and stay focused on slavery. As "*southerners*," he argued, they could "do more at convincing the North than twenty *northern* females," an advantage that they lost by pursuing "*another* subject." He thought that "The *great* reason why *you* should operate upon the public mind far and wide at the north rather than Mrs. Child, Mrs. Chapman, Lucretia Mott, etc. is that you are *southern* women, *once* in *law* slaveholders, your friends all slaveholders, etc., hence your testimony; *testimony* TESTIMONY is the great desideratum."[64] (See Document 26.) John Greenleaf Whittier, a Quaker poet and abolitionist writer, seconded Weld's view, supporting the sisters' public speaking on women's rights, but urging them not "to enter the lists as controversial writers on this question." (See Document 27.)

Standing her ground, Angelina explained to both men why she and Sarah felt they now had to defend women's rights in their speaking and writing. "If we surrender the right to *speak* to the public this year, we must surrender the right to petition next year & the right to write the year after &c.," she wrote. "What *then* can *woman* do for the slave, when she herself is under the feet of man & shamed into *silence?*" (See Document 28.)

The annual reports of many women's antislavery societies endorsed the Grimkés' fight for women's rights. At an October 1837 meeting the Ladies Anti-Slavery Society in Providence, Rhode Island, resolved "that we act as moral agents" and "that our rights are sacred and immutable, and founded on the liberty of the gospel, that great emancipation act for women." (See Document 29.) A majority of the members of the nation's largest and most powerful women's abolitionist group, the Philadelphia Female Anti-Slavery Society, also voted their approval. (See Document 30.)

[64]Theodore Weld to Sarah and Angelina Grimké, Hartford, Conn., May 22, 1837, in Barnes and Dumond, *Weld-Grimké Letters,* 1: 389. For Child, Chapman, and Mott, see Hersh, *Slavery of Sex.*

The sisters had another reason for insisting that they needed to write about women's rights. At this early moment in the discussion of women's rights, people did not know how to "sustain their ground by argument." (See Document 28.) In a burst of writings in 1837 and 1838, Angelina and Sarah Grimké meant to provide such arguments.

In October 1837, the sisters settled into the home of friends in Brookline, Massachusetts. "Oh how delightful it was to stretch my weary limbs on a bed of ease, and roll off from my mind all the heavy responsibilities which had so long pressed upon it," Angelina wrote Jane Smith.[65] She and Sarah had begun to turn their energies to writing about women's rights as early as June, but their stream of writings in the summer of 1837 became a river that autumn.

Both wrote essays in the form of letters that were later published as books, Sarah's as *Letters on the Equality of the Sexes and the Condition of Woman, Addressed to Mary S. Parker, President of the Boston Female Anti-Slavery Society* (1838); Angelina's as *Letters to Catherine E. Beecher, in Reply to An Essay on Slavery and Abolitionism* (1838).[66] Their writings became the standard for women's rights thinking until a decade later, when the Seneca Falls Women's Rights Convention sparked another outpouring of commentary. Glowing with a white-hot radiance on the question, Angelina and Sarah left an enduring legacy for women's rights advocates.

The sisters' writings on women's rights relied primarily on religious arguments. Yet because they buttressed these arguments with Enlightenment notions about human equality and natural rights contained in the Declaration of Independence and the Bill of Rights, their vision of female equality extended further than the "Biblical Feminism" of itinerant female preachers. In these "letters" they explored the implications of the ideas they had developed during their speaking tour. Their arguments had three major dimensions. First, they insisted that all rights were grounded in "moral nature." Second, they explored the personal aspects of moral identity and of rights. Third, they analyzed the social and political meaning of women's rights.

[65] AG to Jane Smith, Holliston, Mass., Oct. 26, 1837, Clements Library.

[66] Sarah M. Grimké, *Letters on the Equality of the Sexes, and the Condition of Woman, Addressed to Mary S. Parker, President of the Boston Female Anti-Slavery Society* (Boston: Knapp, 1838); and Angelina E. Grimké, *Letters to Catherine E. Beecher, in Reply to An Essay on Slavery and Abolitionism, Addressed to A. E. Grimké. Revised by the author.* (Boston: Knapp, 1838). The versions reprinted here are those that originally appeared in *The Liberator.* These versions have the virtue of the immediacy of the moment; Theodore Weld later aided in the revisions that appeared in the book. See also Sarah Grimké, *Letters on the Equality of the Sexes and Other Essays,* ed. Elizabeth Ann Bartlett, (New Haven: Yale University Press, 1988).

Angelina's twelfth letter to Catharine Beecher was in many ways the most eloquent of the sisters' writings on women's rights. (See Document 31.) She began with a tribute to the abolition movement as a school that taught her about women's rights:

> Since I engaged in the investigation of the rights of the slave, I have necessarily been led to a better understanding of my own; for I have found the Anti-Slavery cause to be the high school of morals in our land — the school in which human rights are more fully investigated, and better understood and taught, than in any other benevolent enterprise.

That "investigation" led Angelina to understand the parallels between the slave's lack of liberty and her own lack of liberty as a white woman.

Perhaps even more important, the antislavery cause encouraged Angelina Grimké to formulate one principle capacious enough to sustain equal rights for women and men, slaves and masters. Her twelfth letter explored that "one great fundamental principle,"— that "the rights of all men, from the king to the slave, are built upon their moral nature: and as all men have this moral nature, so all men have essentially the same rights." These rights might be "plundered," she said, "but they cannot be alienated." If rights were "founded in moral being," she insisted, "then the circumstances of sex could not give to man higher rights and responsibilities, than to woman." Those who argued that it did, she thought, denied "the self-evident truth, 'that the physical constitution is the mere instrument of the moral nature.'"

Angelina's argument was based on her interpretation of the equality of men and women at the moment of their creation. "I affirm that woman never was given to man. She was created, like him, in the image of God, and crowned with glory and honor; created only a little lower than the angels,— not, as is too generally presumed, a little lower than man." Angelina gave her readers a new image of women as endowed with the symbols of divine authority. "On her brow, as well as on his, was placed the 'diadem of beauty,' and in her hand the scepter of universal dominion." The radical simplicity of Angelina's argument allowed her to ignore the difficulties that women encountered when they tried to construct feminist interpretations of scripture.[67] Through her reading of the creation

[67] Sarah's *Letters on the Equality of the Sexes* struggled more explicitly with scripture. Her efforts inspired a more thorough work, Elizabeth Wilson, *A Scriptural View of Woman's Rights and Duties in All the Important Relations of Life* (Philadelphia: Young, 1849). For an appraisal of biblical criticism by SG and Lucretia Mott, see Marla J. Selvidge, *Notorious Voices: Feminist Biblical Interpretation, 1500–1920* (New York: Continuum, 1996), 44–62;

story Angelina gave her contemporaries a powerful new myth that naturalized and legitimized women's equality and denaturalized and delegitimized male superiority.

Upon this new reading of the creation story Angelina built her argument that women should have equal rights in the secular social and political world. When she supported woman's "right to be consulted in all the laws and regulations by which she is to be governed, whether in Church or State," she was making a very large claim indeed. In a passage with closing words that must have thrilled some readers and shocked others, she further insisted "that woman has just as much right to sit in solemn counsel in Conventions, Conferences, Associations, and General Assemblies, as man — just as much right to sit upon the throne of England, or in the Presidential chair of the United States, as man."

"The fundamental principle of moral being" was a flexible concept that allowed Angelina to see two dimensions in women's equality with men: in some ways they were the same as men, and in some ways they were different. Urging the similarity between women and men, her twelfth letter denounced "the anti-christian doctrine of masculine and feminine virtues." But this did not prevent her from arguing, as she did throughout her public years, that women's social circumstances created moral duties that were particular to women. "Where *woman's* heart is bleeding, shall *woman's* heart be hushed?" remained for her a compelling question. (See Document 7.) In her last public speech she mentioned her special duties as a southerner. So too women had special duties. In 1838 she appealed to the "women of Philadelphia" as women. Since they had no right to vote, she thought "it is . . . peculiarly your duty to petition." (See Document 34.)

The Grimkés' concept of "moral being" did not mean that women had to be the same as men in order to be equal to men. "Moral being" was a concept that recognized women's circumstances as historically and socially constructed. This view helped them recognize women's claim on society as a group. The Grimkés' view of women's rights did not prompt her to see women as interchangeable units shorn of social context, but as collective beings with moral voices capable of expressing their collective experience. Viewed from this perspective, their writings have a great deal to offer current debates over feminist theory. "Moral being" is a concept that can sustain arguments on behalf of affirmative action (a collective treatment of women that tries to promote women's

and Cullen Murphy, *The Word According to Eve: Women and the Bible in Ancient Times and Our Own* (Boston: Houghton Mifflin, 1998), 25–29.

equality with men by recognizing their socially constructed differences from men) as well as equal rights (their right to be treated as equal human beings).[68]

In her *Letters on the Equality of the Sexes,* Sarah Grimké also based her arguments on the equality of men's and women's "immortal being." She offered scathing social criticisms on a topic of compelling interest to all women — the institution of marriage. In her twelfth letter — "Legal Disabilities of Women" — Sarah drew parallels between the legal status of married women and that of slaves. Women were deprived of "responsibility . . . as a moral being, or a free-agent" by laws that submerged their "very being" into their husbands. The legal system empowered men to "chastise" women physically, to deprive them of their property and their earnings. These laws also reinforced men's power by lessening women's estimation of themselves "as moral and responsible beings," teaching them "to look unto man for protection and indulgence." (See Document 32.)

In addition to the legal system, however, personal relations between husband and wife were shaped by scriptural authority that required wives to submit to their husbands. For nineteenth-century women this often meant submitting to unwanted sexual intercourse, and since intercourse often led to pregnancy, to unwanted pregnancy. Indirectly addressing this vital issue, Sarah's thirteenth letter offered a new interpretation of the biblical admonition, "Wives, submit yourselves unto your own husbands."[69] The Bible also required husbands to honor their wives, she said. Moreover, "submit" and "subjection" were terms that recommended a Christian spirit of humility, not a literal rule of husbands over wives. (See Document 33.)

After writing the last of their essays in the winter of 1838, Angelina and Sarah ended their careers as public speakers with a series of spectacular events. In January they participated in a debate at the Boston Lyceum on the question, "Would the condition of woman and of society be improved by placing the two sexes on an equality in respect to civil rights and duties?" In February Angelina became the first woman to address the Massachusetts state legislature, speaking in Representatives' Hall on the slave trade in the District of Columbia, and the interstate slave trade. "We Abolition Women are turning the world upside down,"

[68] For a guide to recent debates within feminist theory, see Mary Lyndon Shanley and Uma Narayan, *Reconstructing Political Theory: Feminist Perspectives* (University Park: University of Pennsylvania Press, 1995).

[69] Ephesians 5:22.

Angelina wrote Sarah Douglass.[70] In March and April under the auspices of the Boston Female Anti-Slavery Society they offered a series of lectures, the first of which was attended by 2,800 people.

In early February 1838, Theodore Weld declared his love and proposed marriage to Angelina Grimké. To gain her consent he wrote only one letter, in which he said that her 1835 letter to Garrison "formed an era in my feelings and a crisis in my history that drew my spirit toward yours." Her response revealed how passionately she felt about that turning point in her life. She attributed insights to him that he may not have had:

> You speak of my letter to W. L. G. Ah! you felt then that it was written under tremendous pressure of feelings bursting up with volcanic violence from the bottom of my soul—you felt that it was the first long breath of *liberty* which my imprisoned spirit dared to respire whilst it pined in hopeless bondage, panting after freedom to *think aloud.*[71]

She agreed to marry him, saying that she loved him as "a kindred mind, a congenial soul" with whom she "longed to hold communion."[72] They set a wedding date for May.

The "volcanic violence" of Angelina Grimké's commitment to Garrisonian abolitionism helped her break customary limits on women's participation in public life. By thinking aloud—fearlessly—she and her sister brought new words and new concepts into public life.

Perhaps reflecting the burdens of her pathbreaking three years, Angelina retired from public life after her marriage to Theodore Weld, and Sarah joined the couple in their rural New Jersey retreat. Theodore, the most charismatic male speaker employed by the AASS, had lost his voice after one particularly exhausting lecture tour in 1836, and never fully recovered it. Theodore supported his growing family by becoming a teacher, although he continued to play a part in the antislavery movement. For example, he published in 1839 *American Slavery as It Is,* a scathing documentary indictment of slavery drawn from personal narratives (including those written by Angelina and Sarah) and from Southern newspapers published between 1837 and 1839.

[70] AG to Sarah Douglass, Brooklyn (Mass.), Feb. 25, 1838, in Barnes and Dumond, *Weld-Grimké Letters,* 1: 574.

[71] Theodore Weld to AG, New York, Feb. 8, 1838, and AG to Theodore Weld, Brookline (Mass.), Feb. 11, 1838, in Barnes and Dumond, *Weld-Grimké Letters,* 1: 533–34 and 536–37.

[72] AG to TW, Feb. 11, 1838, Barnes and Dumond, *Weld-Grimké Letters,* 2: 536–38.

THE ANTISLAVERY MOVEMENT SPLITS OVER THE QUESTION OF WOMEN'S RIGHTS, 1837–1840

Just after her wedding in May 1838, Angelina Grimké spoke in public for the last time before the second convention of antislavery women in Philadelphia. Her lecture, recorded by a stenographer, shows how she stood up to the pressure of a stone-throwing mob estimated to contain about 10,000 men, who were breaking the windows of the hall in which she spoke. (See Document 34.) The convention was meeting in newly completed Pennsylvania Hall, a building constructed by Garrisonian abolitionists and other radical groups who were often denied access to the city's other public buildings, especially its churches. The mob was especially enraged by the interracial character of the women's convention. While some black men attended AASS meetings, a higher proportion of participants in the women's convention were black women. In the midst of the din, the mayor asked the women to adjourn. They left the hall walking arm in arm, two and three abreast, black women protected by white women on each side, into the multitude of "fellows of the baser sort" who parted to make a narrow path. The next night, after breaking into the building and smashing its furnishings, the mob burned Pennsylvania Hall to the ground.[73] (See Figure 7.)

Angelina's antislavery career ended as it had begun three years earlier — amid the turbulence of mob violence. That violence symbolized the intractability of slavery as an issue in U.S. society, and it highlighted the degree to which Garrisonians were challenging the assumptions on which slavery rested. However, this adamant rejection of their message led many abolitionists to conclude that Garrisonian moral suasion was not working, and that other strategies must be tried.

The Garrisonian movement itself was changing. In addition to women's rights, other reforms, most notably perfectionist views of pacificism and opposition to government, were becoming part of the movement. In 1837 Garrison himself began to express perfectionist views in *The Liberator.* He added "Universal Emancipation" to the paper's old motto, "Our Country is the world — our countrymen are all mankind." (See Figure 8.) Henceforth, he said, the paper would support "emancipation of our whole race from the dominion of man, from the thraldom

[73] ***History of Pennsylvania Hall, Which Was Destroyed by a Mob, on the 17th of May, 1838*** (Philadelphia: Merrihew and Gunn, 1838; reprint, New York: Negro University Press, 1969), 135. Lerner, *The Grimké Sisters,* 247.

Figure 7 The burning of Pennsylvania Hall, May 17, 1838. Constructed in 1837 by the donations from the city's radical reform organizations, the hall was burned to the ground by a mob of about 10,000 anti-abolitionist men.
'The Burning of Pennsylvania Hall,' Courtesy the Quaker Collection, Haverford College Library.

FREEDOM ST.

EMANCIPATION

THE LIBERATOR

OUR COUNTRY IS THE WORLD, OUR COUNTRYMEN ARE ALL MANKIND.

Figure 8 Masthead of *The Liberator*. In March 1838, *The Liberator* added a second panel (on the right) to its masthead. The first masthead (panel on the left) depicted humans being sold with horses and cattle. The new panel evoked universal emancipation and a bright, industrious future.
Boston Athenæum.

of self, from the government of brute force, from the bondage of sin." Garrison's new views on government were especially radical.

> As to the governments of this world . . . we shall endeavor to prove, that, in their essential elements, and as presently administered, they are all Anti-Christ; that they can never, by human wisdom, be brought into conformity to the will of God; that they cannot be maintained except by naval and military power.[74]

These ideas became the basis of the New England Non-Resistant Society, founded by Garrison and others in 1838. Springing from a religious quest for perfect holiness to which Garrison and many of his supporters rededicated themselves in 1838, these views of government also arose from their view of slavery as a form of force and of official tolerance of that force by secular and clerical authorities as particularly pernicious. As in the 1960s, when it was expressed by Martin Luther King, Jr. as nonviolent civil disobedience, this stance expressed engagement rather than retreat from the world. But in the 1830s it more radically rejected human institutions as a means of achieving human freedom.

Angelina approved of Garrison's move, writing Henry Clarke Wright in August 1837, "What wouldst thou think of the Liberator abandoning Abolition as a primary object and becoming the vehicle for all these grand principles? Is not the time rapidly coming for such a change. . . . O how lamentably superficial we are, to suppose that one truth can hurt another; as well might we suppose that to teach one branch of science is to undermine another."[75] Angelina also wrote approvingly to Jane Smith about Wright's perfectionism. "I am truly glad to find that Brother Wright has been with you long enough to explain his ultra peace views. . . . How terrible must be the shaking which will shake down the vast structures which man has built up to fetter the mind & body of his fellow man."[76]

Between 1838 and 1840 these changes within the Garrisonian movement generated internal conflict that split the movement. Many abolitionists — men and women — feared that Garrison's promotion of a wide range of perfectionist views brought disrepute to the antislavery movement. Certainly these views made it difficult for his associates to use political institutions to combat slavery. In 1839–40 the movement split into three branches. Garrison's supporters maintained control of the AASS; and after 1839 women exercised even greater power within the society.

[74] *The Liberator,* Dec. 15, 1837.

[75] AG to Henry Clarke Wright, Brookline (Mass.), Aug. 27, 1837, in Barnes and Dumond, *Weld-Grimké Letters,* 1: 436.

[76] AG to Jane Smith, Brookline (Mass.), Aug. 26, 1837, in Ceplair, *The Public Years,* 285.

Garrison's opponents formed two new groups. Some created the American and Foreign Anti-Slavery Society (A&FASS) in 1839, excluding women from membership and relegating them to auxiliary groups. (Both its supporters and opponents called this "the new organization.") Others put their energies into politics and in 1840 formed the Liberty Party, within which women played a marginal role as nonvoters, though a significant role as fundraisers.[77]

While Garrison's critics mentioned his perfectionist views, they justified their separation from the AASS on the grounds that Garrisonians were insisting on the advancement of women within the organization. The precipitating cause was the appointment of Abby Kelley to the AASS business committee in 1840. (See Document 35.) Kelley, a thirty-year-old Quaker teacher from Lynn and strong supporter of the Grimkés, first spoke in public at the Second Women's Anti-Slavery Convention, joined Garrison in founding the New England Non-Resistant Society in 1838, and in 1839 became an AASS lecturer. She was the first woman appointed to a governing committee of the AASS. (See Figure 9.)

Officers of the new organization blamed AASS members for the split, since they had brought "females into the business meetings to vote and speak, and also that they should be appointed officers of the Society." If the word "person" in the AASS Constitution was meant to include women, they argued, "then, it must, of course, include, not only women, but children also." The AASS Executive Committee decried the new organization's limitation of membership rights to men. They insisted that the AASS did not require members to believe in the equality of the sexes, nor adopt ultraist views of government. However, the AASS did refuse "to dictate to abolitionists" on these topics or repress unpopular views. This, they said, was what the separatists wanted.[78]

Angelina and Sarah Grimké viewed the fracturing of the abolitionist movement from their domestic retreat. In a world in which unmarried women were almost never able to maintain their own homes, Sarah lived with Angelina and Theodore for the rest of her life. All three were determined to show that Angelina and Sarah's advocacy of women's rights had not made them unfit for family life. "Now I verily believe that we are *thus* doing *as much* for the cause of woman as we did by public speaking," she

[77] On the movement's split, see Aileen S. Kraditor, *Means and Ends in American Abolitionism: Garrison and His Critics on Strategy and Tactics, 1834–1850* (New York: Random House, 1967); and Lewis Perry, *Radical Abolitionism: Anarchy and the Government of God in Antislavery Thought* (Knoxville: University of Tennessee Press, 1995).

[78] "Address to the American and Foreign A.S. Society," *The Liberator,* June 19, 1840; "Address of the Executive Committee of the American Anti-Slavery Society," *The Liberator,* July 31, 1840.

Figure 9 Abby Kelley at age 41 in 1851. Theodore Weld recruited Kelley to become an antislavery lecturer after her powerful speech at the women's antislavery convention in Philadelphia, May 1838.
Reprinted from Dorothy Sterling, *Ahead of Her Time: Abby Kelley and the Politics of Antislavery* (1991). Courtesy of Dorothy Sterling.

wrote Anne Weston in the summer of 1838. "For it is absolutely necessary that we should show that we are *not* ruined as domestic characters." (See Document 36.) Angelina's conviction was shaped by the combination of her notorious prominence in introducing women's rights into antislavery discourse, by her deviant status as a single woman at the age of 33 in 1838, and by her hope to make a contribution to the reform of family life. Given the first two realities, her hope for the third required her to prove herself as a wife and mother. In the next six years Angelina gave birth to three children. Other women's rights abolitionists did not share Angelina's fears. Lucretia Mott, married at the age of seventeen in 1811, and the mother of six children by 1828, confidently wove her reform activism together with a gratifying family life. Angelina, having ventured further out in challenging her society's norms in her speaking career, felt a larger obligation to show that women's rights were compatible with family life.

Although Angelina rejoiced in her "release from public service," she maintained a keen interest in the fracture of the abolition movement over the "Woman Question." She was confident that women's rights could "no more be driven back . . . than the doctrine of Human rights, of which it is a part, & a very important part." And she thought that women's rights should become a part of "*every* reform." (See Document 36.)

Other women had different reactions to the antislavery rupture. Lydia Maria Child, in September 1839, provided a detailed interpretation of how and why the movement was splitting into pro- and anti-Garrison factions. Child, married but without children, had sacrificed a comfortable career as a writer for young people when she joined the abolitionist movement in 1833 by writing *An Appeal in Favor of That Class of Americans Called Africans.*[79] In 1840 she was named to the executive committee of the American Anti-Slavery Society and in 1841 became the editor of its weekly, the *National Anti-Slavery Standard,* making her, along with Abby Kelley, the most prominent Garrisonian woman now that the Grimkés had retired. Child explained that she and others had not pushed women's rights: "Instead of forcing this 'foreign topic' into antislavery meetings or papers, we have sedulously avoided it." But when Garrison's critics asked women to withdraw from the AASS, she refused. She thought that clergymen were the main opponents of women leaders within the movement. And she felt "these questions of Non-Resistants, Woman's Rights, &c. are only urged to effect a secret purpose," which was to topple Garrison's leadership. Yet Child was not discouraged, believing that the conflict would eventually "mightily promote the cause of general freedom." (See Document 37.)

All women's antislavery organizations underwent a painful period of reevaluation when the movement split. Many women shared the concern of Garrison's male critics that perfectionism was tainting abolitionism with heretical religious views and unsound political opinions. This

[79] Lydia Maria Child, *Appeal in Favor of That Class of Americans Called Africans* (Boston: Allen and Ticknor, 1833). See also Carolyn L. Karcher, ed., *A Lydia Maria Child Reader* (Durham, N.C.: Duke University Press, 1997).

was especially true of middle-class evangelical women who were also members of moral reform societies. Where that group was strong, as in the Boston, New York, and Rochester societies, those organizations splintered and ceased to exist. Those outside the middle-class mainstream, such as Hicksite Quakers like Amy Post of Rochester, or elite Bostonians like Maria Weston Chapman, supported Garrison and women's rights. The Philadelphia society survived, partly because many of its members were Hicksite Quakers.[80] But the elite women within Boston's society were not numerous enough to save it. In smaller societies throughout the Northeast, women were at best dispirited. "We are very feeble," wrote one Uxbridge, Massachusetts, woman.[81] Although most black men and women remained loyal to Garrisonian organizations, most also regretted the diversion of so much energy from the movement's main goal of ending slavery.

The Boston Female Anti-Slavery Society self-destructed in an acrimonious clash between Garrisonians, who included Lydia Maria Child, and their opponents, led by Mary Parker, the president. Women's rights were not mentioned at this meeting. Instead the dispute pivoted around the "new organization" (the American and Foreign Anti-Slavery Society). Even though the new group opposed the advocacy of women's rights within abolitionism, one woman asked in the heat of the debate, "Have we not the right to sympathize with it?"[82] (See Document 38.)

This crisis became an important catalyst to the growth of women's rights within Garrisonian abolitionism. Other leaders emerged to carry women's rights forward. Chief among these was Abby Kelley, who met opponents with the same logic used by the Grimkés—"Whatever ways and means are right for men to adopt in reforming the world, are right also for women to adopt."[83] A popular speaker in the 1840s, Kelley traveled throughout the Northeast carrying Garrison's message of "No Union with Slaveholders" and his denunciation of the Constitution as a "covenant with death."

In Kelley's wake, dozens of antislavery societies formed. Some, like the Western New York Anti-Slavery Society (WNYASS), were especially

[80] See Carolyn Luverne Williams, "Religion, Race, and Gender in Antebellum American Radicalism: The Philadelphia Anti-Slavery Society, 1833–1870," Ph.D. diss. (UCLA, 1991). For a comparison of the social agendas of moral reform and perfectionist activists, see Nancy A. Hewitt, *Women's Activism and Social Change: Rochester, New York, 1822–1872* (Ithaca: Cornell University Press, 1984).

[81] Quoted in Jeffrey, *The Great Silent Army of Abolitionism*, 98.

[82] For more on the Boston society, see Hansen, *Strained Sisterhood.*

[83] *Connecticut Observer,* March 7, 1840, 38; quoted in Keith Melder, "Abby Kelley Foster," *Notable American Women.* See also Sterling, *Ahead of Her Time.*

vocal in advocating women's rights as well as emancipation. The WNYASS regularly elected women to offices in the society and sent women as delegates to the annual conventions of the AASS. Kelley drew other talented women to the Garrisonian ranks, including Paulina Wright Davis, Lucy Stone, and Susan B. Anthony. In 1845 Kelley helped found and thereafter raised the funds necessary to keep alive *The Anti-Slavery Bugle* in Salem, Ohio. That year she married Stephen Foster, another Garrisonian radical, and although she became a mother in 1847, unlike the Grimkés, she continued her lecture career while her husband cared for their child. In the mid-1850s Kelley broke with Garrison. Both thought that the time had come for abolitionists to become involved in politics, but Garrison supported the new Republican party and Kelley Foster wanted an abolitionist party.

Kelley's career demonstrated both the opportunities and the limitations of women's rights activism within the Garrisonian movement. Because Garrisonian radicalism became a seedbed for the cultivation of radical ideas in the 1830s, women's rights could emerge and flourish there. Yet Garrisonian radicalism was on a trajectory that carried it away from the mainstream of American public culture, and its boldest adherents spent increasing time in the 1840s and 1850s trying to maintain the movement's centrality in the face of other forms of antislavery activism. Thus Abby Kelley attracted important recruits to the cause, but she produced no writings that advanced women's rights beyond the achievements of the Grimkés. Nevertheless her recruits to the antislavery cause embraced the Grimkés' dual agenda of women's rights and antislavery, and in the 1850s they formed the leadership base for a new movement dedicated to women's rights in the United States.

In this context, Garrisonian abolitionism no longer served as the most promising soil for the continued growth of women's rights. The movement was too often at odds with the mainstream of middle-class political culture. To help women's rights flourish, its advocates had to change it. Above all, they had to find a new location for it in American political culture.

AN INDEPENDENT WOMEN'S RIGHTS MOVEMENT IS BORN, 1840–1858

Two women led the effort to transplant women's rights into a new reform environment between 1840 and 1848. One, Lucretia Mott (1793–1880), a senior stateswoman in the Garrison movement, represented the radical

Figure 10 Lucretia Mott at about age 49 and James Mott, c. 1842. Like the Grimké sisters in the 1830s and Stanton and Anthony after 1852, the Motts constituted another set of reformers whose public power was increased by their mutual support.
Courtesy of the Massachusetts Historical Society.

religious impulse that had sustained the Grimkés' innovations. (See Figure 10.) The other, Elizabeth Cady Stanton (1815–1902), represented a younger generation who viewed women's rights as a secular issue. (See Figure 11.) In 1840 Elizabeth Cady married Henry Stanton, a leading member of the "new organization" (the American and Foreign Anti-Slavery Society), but she remained independent of the antislavery movement and its internal differences. In July 1848, Mott and Stanton orga-

Figure 11 Elizabeth Cady Stanton, c. 1848, a determined young matron with her sons Henry and Daniel.
By permission of Rhoda Jenkins/Coline Jenkins-Sahlin.

nized the enormously successful Women's Rights Convention in Seneca Falls, New York, and authored the convention's historic "Declaration of Sentiments." (See Document 42.) That event launched a series of women's rights conventions during the 1850s, which, after an interregnum during the U.S. Civil War in the 1860s, led in 1869 to the founding of national and local women's suffrage organizations, and eventually to the adoption of the Nineteenth Amendment to the U.S. Constitution.

To her collaboration with Stanton, Mott brought her considerable oratorical talents and her reputation as a leading abolitionist and a defender of women's rights within Hicksite Quakerism. She became the moving spirit of the 1848 convention and attracted dozens of Hicksite supporters to the new movement. Stanton brought from her father's law practice an understanding of the way that laws empowered men and disadvantaged women. She kept their efforts focused on legal issues. Together Mott and Stanton rescued women's rights from the sectarian limitations of the Garrisonian movement. At the same time, however, they also removed women's rights from many of the religious and spiritual moorings to which the Grimkés had attached it.

Just as the Grimkés' destinies unfolded from the unexpected death of their father in Philadelphia, so the friendship of Lucretia Mott and Elizabeth Stanton flowed from their chance encounter in London in the summer of 1840. There they both attended the World's Anti-Slavery Convention, organized by the British and Foreign Anti-Slavery Society. Mott was a delegate from Pennsylvania, representing both the Philadelphia Female Anti-Slavery Society and the Pennsylvania Anti-Slavery Society. Stanton had no official capacity; she was accompanying her husband, a delegate representing the "new organization." Elizabeth Cady and Henry Stanton had married in May; their London trip was a honeymoon. Elizabeth was apparently highly motivated to attend the convention. After Henry Stanton told her he would be attending the event, she agreed to marry him, putting aside her father's objections to Stanton's dim financial prospects and her own previous doubts about the match.[84]

The friendship that Elizabeth Stanton formed with Lucretia Mott during their three weeks in London marked a turning point in the younger woman's life. The gender politics of the convention, the chemistry of her relationship with Mott, and her own personal agenda prompted her to view Mott as a model and a mentor. "Mrs. Mott was to me an entirely new revelation of womanhood," Stanton later wrote. (See Document 39.)

The host organization, the British and Foreign Anti-Slavery Society, supported the "new organization" in its separation from the Garrisonian wing of abolitionism, and demonstrated that support by refusing to seat the seven women delegates from the United States. No male Garrisonian delegate was refused a seat. Although the hosts never publicly stated their reasons for excluding the women, delegates were told "that it would be outraging the tastes, habits, customs and prejudices of the English people, to allow women to sit in this Convention."[85] Lucretia Mott

[84]Theodore Stanton and Harriot Stanton Blatch, eds., ***Elizabeth Cady Stanton, As Revealed in Her Letters, Diary and Reminiscences*** (New York: Harper & Bros., 1922), 67.

[85]"Proceedings of the General Anti-Slavery Convention," quoted in Kathryn Kish Sklar, "'Women Who Speak for an Entire Nation': American and British Women at the World Anti-Slavery Convention, London, 1840," in *The Abolitionist Sisterhood: Women's Political*

and other women delegates were surprised to discover that British women abolitionists, many of whom were Quakers, did not support the seating of U.S. women delegates.[86]

This difference in the gender politics of the American and British movements arose in part from differences in the nature of the struggle against slavery in the two countries. Slavery in the British West Indies was abolished by an act of Parliament in 1833. In the United States, however, it took one of the bloodiest wars in human history to end slavery a generation later. To oppose slavery in the United States in the 1830s on terms other than the ineffective proposals of the Colonization Society required a radical stance capable of challenging the nation's social, political, and economic status quo. In that context it became easier to challenge accepted gender relations. British abolition did not open up the same possibilities. Although some British women abolitionists were attracted to American notions of gender equality, their abolitionist movement did not become the seedbed for a women's rights movement. British women were not admitted to membership in the British and Foreign Society; they joined female auxiliaries and were represented by men in the men's group. England had no equivalent to Angelina Grimké, Lucretia Mott, or Abby Kelley.

Another source of the different views of British and American women on the question of seating women at the 1840 convention lay in the nearly unanimous effort among British Quakers to seek acceptance in a society that limited the right to vote to members of the Church of England. In their quest for the right to vote, British Quakers had distanced themselves from their radical origins and were adopting many forms of Anglican worship, including creeds that upheld the divinity of Jesus. This trend had also affected American Quakers, producing the schism of 1827, which divided them into Orthodox and Hicksite groups. Having no equivalent to Hicksites in England, British Quakers viewed Lucretia Mott as a religious heretic, and most British women abolitionists avoided meeting with her. Mott's more traditional form of Quakerism, which disavowed all authority except individual conscience, threatened the respectable reputation that British Quakers were trying to construct. "I fear thy influence on my children!!" one told her. (See Document 39.)

Elizabeth Stanton found this drama centered around Lucretia Mott intensely interesting. Writing about the occasion forty years later, she

Culture in Antebellum America, ed. Jean Fagan Yellin and John C. Van Horne (Ithaca: Cornell University Press, 1994), 311.

[86]For the convention, see Sklar, "Women Who Speak for an Entire Nation."

said that although she had been thoroughly indoctrinated on the ocean voyage by Liberty party men about "the women who had fanned the flames of dissension, and had completely demoralized the Anti-Slavery ranks," she found herself agreeing with the women delegates. At dinner one evening early in the convention, she sat next to Lucretia Mott and observed her skillful defense of women's rights against "most of the gentlemen at the table." "Mrs. Mott parried all their attacks, now by her quiet humor turning the laugh on them, and then by her earnestness and dignity silencing their ridicule and sneers." In the midst of this colloquy, the two women bonded. Elizabeth wrote: "I shall never forget the look of recognition she gave me when she saw, by my remarks, that I comprehended the problem of woman's rights and wrongs. How beautiful she looked to me that day." (See Document 39.) In her diary Lucretia Mott returned the compliment, writing, "Elizabeth Stanton gaining daily in our affections."[87]

Although they came from quite different backgrounds, and Mott was a generation older, the two women agreed on the importance of Mary Wollstonecraft's 1792 book, *Vindication of the Rights of Women.* Stanton had read it before coming to London, and Mott mentioned Wollstonecraft in one of her London talks. Wollstonecraft's unorthodox personal life meant that this first great English feminist was controversial in her own time, and she remained so in 1840. (Before her marriage to William Godwin in 1797, Wollstonecraft had condemned marriage and was not married when her first child was born.) In recognizing her legacy, Mott and Stanton affiliated with a leading figure of the British and French Enlightenment who advocated rights for women in an age dominated by the expansion of rights to men.[88]

Yet for Elizabeth Stanton, Lucretia Mott in London represented more than an accomplished defender of women's rights. From her she derived two other crucial lessons. She learned that women's rights did not need to be anchored in religious beliefs. And she learned that the advocacy of women's rights was compatible with her position in respectable society.

In London, Stanton drew deeply on Mott's religious counsel. "I sought every opportunity to be at her side, and continually plied her with ques-

[87] Frederick B. Tolles, ed., "Slavery and 'The Woman Question': Lucretia Mott's Diary of Her Visit to Great Britain to Attend the World's Anti-Slavery Convention of 1840," *Journal of the Friends Historical Society,* suppl. 23 (1952): 41.

[88] Elizabeth Cady Stanton, Susan B. Anthony, and Matilda Joslyn Gage, *History of Woman Suffrage, 1: 1848–1861* (New York: Fowler & Wells, 1881), 421, 423–24. See also Moira Ferguson, "Mary Wollstonecraft and the Problematic of Slavery," in *Mary Wollstonecraft and 200 Years of Feminisms,* ed. Eileen Janes Yeo (London: Rivers Oram, 1997), 89–103.

tions," she wrote. When Stanton heard her "preach earnestly and impressively" in a Unitarian church, it was the first time she had heard a woman speak in public, which seemed to her "like the realization of an oft-repeated, happy dream." Stanton absorbed Mott's lesson that nothing was too sacred to question, "as to its rightfulness in principle and practice"; that she had "the same right to be guided by my own convictions" as Luther, Calvin, and Knox had to be guided by theirs. On thorny questions of theology, Mott told her, "No one knows any more of what lies beyond our sphere of action than thou and I, and we know nothing." Everything she said seemed "so true and rational," Stanton wrote. "I felt at once a new-born sense of dignity and freedom; it was like suddenly coming into the rays of the noon-day sun, after wandering with a rushlight in the caves of the earth."

In addition to providing liberating religious advice, Mott demonstrated by her example that one could advocate women's rights and still move comfortably in respectable society. Observing Mott at the elegant home of "a rich Quaker banker," Stanton thought she "proved herself in manner and conversation the peer of the first woman in England." (See Document 39.) Mott introduced Stanton to a new dispensation in women's rights — freedom from the Grimkés' religious fervor, while retaining the sisters' radical vision of sexual equality.

Elizabeth Stanton's upbringing meant that she came to reform culture as an outsider. As the cousin of Gerrit Smith, a wealthy abolitionist from upstate New York who founded the Liberty party in 1840, she brought abolitionist credentials to her union with Henry. Furthermore, her privileged girlhood had been shaped by the successful legal practice of her father, Daniel Cady, a prominent lawyer who served in the state legislature, in Congress, and as a judge on the supreme court of New York. In her later writings she said that her father's legal practice, and her acquaintance with his female clients, made her painfully aware of how the law discriminated against women. Contrary to convention, the word "obey" was omitted from her wedding vows, possibly at Henry Stanton's as well as her suggestion, since Henry was a close friend of Theodore Weld and Angelina Grimké, who also omitted "obey" from their ceremony.[89]

Yet Elizabeth did not know what sexual equality looked like in everyday life. Perhaps for that reason she and Henry visited the Weld-Grimké household in New Jersey soon after their wedding. If she had hoped to

[89]See Elisabeth Griffith, *In Her Own Right: The Life of Elizabeth Cady Stanton* (New York: Oxford University Press, 1984), 33.

find there a model for her own married life, she was disappointed. Henry's ties with the trio caused them to expect a comfortable welcome. Henry had been one of the Lane Seminary rebels who had gone to Oberlin from Cincinnati with Theodore Weld. He had served as an agent of the AASS, worked closely with the sisters during their 1837 tour, and supported women's rights. (See Document 24.) A long ride on a cool evening in May had made them hope for "something hot and stimulating," Elizabeth later recalled. Instead, because the household had adopted a health-conscious "Graham diet," they were served "cold dishes without a whiff of heat, or steam, [which] gave one a feeling of strangeness." She remembered: "We were thrown on our own resources, and memories of tea and coffee for stimulus." Elizabeth thought the Weld-Grimké household was "destitute of all tasteful, womanly touches, and though neat and orderly, had a cheerless atmosphere."[90] The sisters had the zeal of converts, which served them well in the crucible of innovation between 1836 and 1838, but despite Stanton's interest in women's rights, Angelina and Sarah Grimké did not inspire imitation in 1840. Lucretia Mott, by contrast, offered the example of a woman who was comfortable with God, with respectable society, and with women's rights. She became the model that Stanton tried to emulate.

In a letter to the Weld-Grimkés about the London convention, Elizabeth Stanton mentioned the discord over "the woman's right question" and her disappointment with the "folly" in an informal speech by William Lloyd Garrison, but her most emphatic words about women's rights were quotations from Lucretia Mott. She passed on a message to both sisters to "write for the public or speak out for *oppressed woman.*" Mott said that "a great struggle is at hand & that all the friends of freedom for woman must rally round the *Garrison standard.*" With her negative estimate of Garrison, Stanton did not seem likely to follow Mott's advice on that score, but Mott's discussion of the "great struggle" must have made her wonder how she could engage in it. (See Document 40.)

Although documents at the time are silent on the matter, Stanton and Mott later said that in London they discussed "the propriety of holding a woman's convention." (See Document 41.) Stanton "made her debut in public" by speaking on temperance in Seneca Falls in the fall of 1841. Mott followed her progress, writing their mutual friends in Dublin that Stanton had moved herself to tears and infused into her talk a "homeopathic dose of Woman's Rights." In her talk Stanton also referred to the sexual double standard that made some things "sinful in woman

[90] "Reminiscences by E.C.S.: Angelina Grimké," in Stanton et al., 392.

alone," showing that her audiences probably included female moral reform society members.[91] Meanwhile, though not affiliated with Garrisonian abolition, Stanton regularly read both of the movement's publications, *The Anti-Slavery Standard* and *The Liberator,* which were sent to her by friends, the latter by Sarah Pugh, Lucretia Mott's close friend. She called these periodicals "the only woman's rights food I have for myself & disciples."[92]

In a letter to Elizabeth Pease, an English abolitionist and friend of Lucretia Mott, whom Stanton had met at the London convention, she explained why she disagreed with the Garrisonian policy of shunning government.

> I am not yet fully converted to the doctrine of no human government. I am in favour of political action, & the organization of a third party as the most efficient way of calling forth & directing action. So long as we are to be governed by human laws, I should be unwilling to have the making & administering of those laws left entirely to the selfish & unprincipled part of the community, which would be the case should all our honest men refuse to mingle in political affairs.[93]

In this regard Elizabeth shared the political views of her husband, who spent most of his time organizing and campaigning, first for the Liberty and then for the Free Soil party.

Her concern about the process of being "governed by human laws" was shaped by debates within abolitionism. But her views had consequences for the emerging women's rights agenda. Garrisonian groups were calling for a boycott of elections. The fourth meeting of the New England Non-Resistance Society in 1842 resolved: "That the use of elective franchise, under the Constitution of the United States, involves the sacrifice of our common humanity and the rejection of the gospel of peace."[94] Such a resolution clearly conflicted with the goal of extending the franchise to women. On the other hand, non-Garrisonians also created a dilemma for her. As she wrote a friend in 1841, "I do not go to any business meetings with Henry because I know I would have no voice in

[91] Lucretia Mott to Richard and Hannah Webb, Philadelphia, Feb. 25, 1842, in *James and Lucretia Mott: Life and Letters,* ed. Anna Davis Hallowell (Boston: Houghton Mifflin, 1884), 228.

[92] Elizabeth Cady Stanton [hereafter ECS] to Elizabeth J. Neall, Johnstown, November 26, 1841, in *The Selected Papers of Elizabeth Cady Stanton and Susan B. Anthony, 1: In the School of Anti-Slavery, 1840–1866,* ed. Ann Gordon et al. (New Brunswick, N.J.: Rutgers University Press, 1997), [hereafter *Papers of Stanton and Anthony*] 25.

[93] ECS to Elizabeth Pease, Johnstown, Feb. 12, 1842, *Papers of Stanton and Anthony,* 1: 29–31.

[94] *The Liberator,* Oct. 28, 1842.

those meetings."[95] By weaving her own independent combination of Garrisonian and non-Garrisonian politics, Stanton created the means for making her entrance into the American political arena.

Elizabeth Cady Stanton first entered that arena in 1843 when, during visits to her father's household in Albany, she joined other women, including Ernestine Rose and Paulina Wright Davis, in circulating petitions to the New York Assembly on behalf of a New York Married Women's Property Act.[96] Beginning with Mississippi in 1836, states were passing legislation that permitted wives to retain ownership of property they brought into marriages. New York eventually passed a married women's property act in 1848.[97] Designed to protect married women's property from misuse by wastrel or incompetent husbands, such legislation was highly pertinent to Elizabeth Stanton's circumstances, since her homes in Boston and later in Seneca Falls were purchased by her father rather than her husband. During these years Henry did little to sustain their middle-class standard of living. Taking a dim view of his son-in-law's political enthusiasms, Judge Cady supported the passage of the 1848 law. Until then his daughter lived in houses that were registered in her name but owned by her husband.[98]

When, at Henry's insistence, the Stantons moved from Boston to Seneca Falls in June 1847, they entered the most congenial environment for women's rights in the United States. A decade earlier the effects of the Second Great Awakening in central and western New York had been especially pronounced. The region's yeasty mix of Hicksite Quakers, evangelical perfectionists, and benevolent activists pursued competing agendas that inspired one another to ever greater efforts. Women's rights flourished there in the 1840s. Hicksite Quakers petitioned their chief governing body, the Yearly Meeting, to grant more power to women within local meetings. Evangelical and benevolent women in Rochester's Ladies Washingtonian Total Abstinence Society did more than campaign against drunkenness. In 1846 they collected 1,400 signatures on a petition that bemoaned women's "lack of the ballot." Rochester's Female Moral Reform Society, which claimed 500 members in 1845, mobilized against prostitution and the sexual double standard that condoned male promiscuity.[99]

[95] ECS to Elizabeth Neall, Nov. 26, 1841, in Gordon, *Papers of Stanton and Anthony,* 25.
[96] Stanton, *History of Woman Suffrage,* 1: 63–67.
[97] See Norma Basch, *In the Eyes of the Law: Women, Marriage, and Property in Nineteenth-Century New York* (Ithaca: Cornell University Press, 1982).
[98] Griffith, *In Her Own Right,* 44–49.
[99] Hewitt, *Women's Activism and Social Change,* 107–13.

Coming from Boston, where she had begun to enjoy a rising place among its reform-minded elite, Elizabeth Cady Stanton was unconnected with these upstate New York networks. Thirty-one years old and the mother of three lively boys, aged five, three, and almost two, she was in charge of her household while her husband traveled on behalf of antislavery political parties. That first year in Seneca Falls she fought for her son's lives against recurring bouts of malaria.[100] In July 1848, when Stanton was invited to spend the day with Lucretia Mott, who was in nearby Waterloo attending the yearly New York meeting of Hicksite Quakers, she "poured out . . . the torrent of my long-accumulated discontent with such vehemence and indignation that I stirred myself, as well as the rest of the party to do and dare anything." Her indignation fell on receptive ears. The Yearly Meeting had just rejected a proposal to grant equal power to women and men in local Quaker meetings under its supervision. Mott, her sister, Martha Coffin Wright, and a neighbor, Mary Ann McClintock, issued a call for "a WOMAN'S RIGHTS CONVENTION . . . to discuss the social, civil, and religious condition and rights of woman" to take place ten days hence, and promised that Lucretia Mott would speak.[101] (See Document 41.)

Stanton brought more than indignation to this group of Hicksite women. They were vastly more experienced than she was about reform activism. But her urgent presence made it easier for them to overcome the Quaker prohibition against joining "popular" non-Quaker reform activities. Their women's rights request having been rejected by their Yearly Meeting, their move to a non-Quaker venue made sense, but an all-Quaker group would have been unlikely to take such a step on its own. As it was, this group of experienced Hicksite women and one young non-church member worked together effectively. They perused "the reports of Peace, Temperance, and Anti-Slavery conventions" as models for how they might organize their convention and phrase their resolutions. These all seemed too tame. Then one of the group "took up the Declaration of 1776, and read it aloud with much spirit and emphasis, and it was at once decided to adopt the historic document." Several "well disposed" men consulted books to help compile the grievances. (See Document 41.)

[100] Griffith, *In Her Own Right,* 48–50.

[101] For the Hicksite influence on this planning meeting, see Judith Wellman, "The Seneca Falls Women's Rights Convention: A Study of Social Networks," *Journal of Women's History,* 3, no. 1 (Spring 1991); and Nancy A. Hewitt, "Feminist Friends: Agrarian Quakers and the Emergence of Woman's Rights in America," *Feminist Studies* 12 (Spring 1986): 27–49.

"Crowds in carriages and on foot" responded to the call, filling the Wesleyan chapel to overflowing, in all about three hundred people. Although the organizers had planned to exclude men the first day of the convention, men showed up and were admitted. In fact, James Mott chaired the meeting. The first day those assembled heard the Declaration of Sentiments and the resolutions that Stanton, Mott, McClintock, and Wright had prepared a few days earlier. That evening Lucretia Mott spoke "on the subject of reforms in general." The second day the convention discussed and adopted the Declaration and resolutions. (See Document 42.)

Historian Judith Wellman found that most of those who came to the convention lived nearby, the majority in Seneca Falls and nearby Waterloo. Others hailed from neighboring counties. Most came with relatives — a sister, parent, child, cousin, or spouse — including Stanton and Mott, whose sisters attended. About a quarter of the attendees were Hicksite Quakers.

Some, including Frederick Douglass, well-known abolitionist editor and former slave, came from Rochester. (See Figure 12.) The only African American at the convention, Douglass had moved to Rochester in 1843. He and his fledgling newspaper, *The North Star* (founded 1847), were indebted to the support of Rochester's antislavery women, especially Amy Post, a Hicksite Quaker who also attended the convention. Douglass was a women's rights supporter when he arrived in Seneca Falls; the masthead of *North Star* declared "Right is of no sex — Truth is of no Color."[102]

Only a few conferees were active in national reform networks, but all presumably participated in the reform movements that flourished in the region. The convention's resolutions and Declaration of Sentiments expressed religious, reform, and political themes, sometimes separately, sometimes woven together into new statements about women's rights and welfare.

The 1848 Declaration of Sentiments improvised creatively on the model of the Declaration of Independence. Its first paragraphs established the right of women to "life, liberty and the pursuit of happiness," and to their duty to "throw off" government that abused them. Thereafter the sentiments divide evenly into legal issues, presumably shaped by Stanton, and a variety of other topics, presumably shaped by Mott, in-

[102] *The North Star,* Rochester, N.Y., Dec. 3, 1847. See also Philip S. Foner, ed., *Frederick Douglass on Women's Rights* (Westport, Conn.: Greenwood Press, 1976); and William S. McFeely, *Frederick Douglass* (New York: Simon & Schuster, 1991), 149–56.

Figure 12 Frederick Douglass at age 25, c. 1843, the time of his arrival in Rochester. Though a supporter of immediate and unconditional emancipation, Douglass remained independent of Garrison.
Boston Athenæum.

cluding women's employment, education, and participation in religious and social reform. Some language in the Declaration of Sentiments echoed the Grimkés and the 1837 women's antislavery convention, especially the rejection of men's "right to assign her a sphere of action, when that belongs to her conscience and to her God."

Yet while God and the Creator appeared frequently in the convention's resolutions, secular phrases adopted from the Declaration of Independence and from the eighteenth-century Enlightenment prevailed over religious discourse from the Second Great Awakening. "Nature" and its "great precept" of happiness was the ultimate authority in the convention proceedings.

In this way Mott and Stanton's collaboration succeeded in transplanting women's rights into the secular soil of nineteenth-century belief in human progress. The third resolution exemplified this process. That resolution asserted "that woman is man's equal," which it supported with two propositions: The Creator intended woman to be man's equal, and "the highest good of the race demands that she should be."

The only resolution that was not unanimously adopted was the ninth that supported women's "sacred right to the elective franchise." This was Stanton's idea, not Mott's, and except for the persistent support of Frederick Douglass, might have been voted down. Eventually a majority at the convention agreed that the right to choose rulers and make laws was crucial because it protected all other rights. Resolutions on the "elective franchise" would become standard fare in future women's rights conventions, and pass with little debate.

Press coverage of the Seneca Falls meeting far exceeded Stanton and Mott's greatest expectations. The entire proceedings were published in both major and minor newspapers, and every editor seemed to have an opinion about the event.[103] Building on the success of the Seneca Falls example, other women's rights conventions met in the 1850s in Ohio, Massachusetts, Indiana, Pennsylvania, and New York. Two weeks after Seneca Falls a convention met in Rochester. Stanton and Mott were distressed when their Rochester imitators went further than they thought acceptable by electing a woman, Abigail Bush, to preside. But that "hazardous experiment" did not slow the movement down. In cities like Worcester, Massachusetts, and towns like Salem, Ohio, the basic format of the Seneca Falls convention was followed: Officers were elected; declarations were made; resolutions were discussed and adopted; formal addresses were heard; letters were read from those who could not attend; and delegates responded to press coverage. Men were active participants on most of these occasions, except the convention at Salem, at which only women were permitted to speak.[104]

The women's rights convention movement established a foundation for the creation of national women's suffrage associations in 1869. Conventions advanced public debate on women's rights issues, and they cultivated the commitment of a generation of dedicated women leaders, many of whom also remained active in other reform movements. Elizabeth Cady Stanton exemplified a trend among some to dedicate them-

[103] Press coverage is described in Stanton et al., *History of Woman Suffrage,* 1: 802–04. In the *New York Tribune,* Horace Greeley provided the women's rights conventions with sympathetic newspaper coverage throughout the 1850s.

[104] Stanton et al., *History of Woman Suffrage,* 1: 74–79, 101–70.

selves solely to women's rights. Lucretia Mott wrote her in the fall of 1848, "You are so wedded to this cause that you must expect to act as pioneer in the work."[105] A few women followed Stanton's lead. Most, however, resembled Mott's pattern of continuing to link women's rights with a broader agenda that gave equal or greater weight to abolition.

Perhaps because women's rights did not become a way of life for Mott, Stanton's relationship with her was supplanted in 1852 by a life partnership that she forged with Susan B. Anthony, a younger Quaker woman who never married. Anthony's parents and sister had attended the Rochester convention, where they heard Stanton read the Declaration of Sentiments. After meeting her briefly at an antislavery convention in Seneca Falls, Stanton recruited Anthony to a partnership that lasted fifty years. Stanton wrote speeches that Anthony delivered; Anthony cared for the children while Stanton wrote the speeches.[106]

Why were women's rights conventions so successful in the 1850s antebellum North? Seven reasons stand out. First, the growth of numerous female antislavery societies in the 1830s and 1840s provided a ready audience for women's rights views. Second, women's rights conventions embodied a new secular perspective that appealed to a larger population than the perfectionist Garrisonians could reach. Third, the conventions emphasized married women's property rights, an important issue to farmers and artisans as well as to elite women. Fourth, women's rights supporters carried their message into other social movements, broadening and deepening the culture's familiarity with women's rights issues. Fifth, the conventions raised a big tent that welcomed women from other reform movements, especially temperance and moral reform. Sixth, transformations in women's family and sexual experience made women's rights relevant to the daily lives of most middle-class women. Seventh, the conventions did not challenge racial prejudice to the same degree as the early antislavery movement had, thereby offering a less arduous venue for white women's activism.

For women abolitionists, women's rights was not a new issue. At Seneca Falls and all subsequent women's rights conventions, a substantial number of participants were attracted by the familiarity rather than the novelty of the subject.

[105]Lucretia Mott to ECS, Oct. 3, 1848. Available on microfilm, *The Papers of Elizabeth Cady Stanton and Susan B. Anthony* (Wilmington, Del.: Scholarly Resources, 1991), reel 6, frame 828.

[106]For this relationship, see Ellen Carol DuBois, ed., *Elizabeth Cady Stanton, Susan B. Anthony: Correspondence, Writings, Speeches* (New York: Schocken, 1981); and Kathleen Barry, *Susan B. Anthony: A Biography* (New York: New York University Press, 1988).

Figure 13 The earliest known picture of Sojourner Truth, from her 1850 narrative. Truth supported herself by lecturing and selling her books. Around 1850 she acquired a house in Northampton, Massachusetts, which she shared with her children.

Sojourner Truth (c. 1797–1883) became an important symbol of the continuity between the antislavery and women's rights movements. (See Figure 13.) Born to slave parents on a New York estate, she was named Isabella, and took the name Van Wagener after a family by that name sheltered her when she ran away in 1827, the year before adult slaves were emancipated in New York. As a slave she bore five children by a fellow slave; one died and two were sold away. At an early age she began to have mystical experiences that she identified as the voice of God. During the 1830s she supported herself and two children as a domestic worker in New York City. Responding to voices that told her to take the name "Sojourner Truth" and to travel and preach, she walked through New England and settled in Northampton, Massachusetts, where she joined Garrisonian abolitionists.[107]

Truth made her first documented public speech at the 1850 women's rights convention in Worcester. There she argued for women's reform activism: "Woman set the world wrong by eating the forbidden fruit, and now she was going to set it right." On a perfectionist note, she also said, "Goodness was from everlasting and would never die, while evil had a beginning and must come to an end." In Mott's closing address to the convention, she warmly repeated Truth's words. After staying for a few months with Amy Post in Rochester, Truth headed to Ohio to attend other women's rights conventions, making her headquarters at the offices of the *Anti-Slavery Bugle* in Salem, Ohio.

In her famous speech at the Akron, Ohio, women's rights convention in 1851, Sojourner Truth did not discuss the needs of free black women, but she did demonstrate her own wit and insight based on her experience as a former slave. A large, charismatic person, she boasted about her physical strength, "I am as strong as any man that is now." She deferred to men as having greater intellect, but still insisted on women's right to their own portion. Truth gave her own interpretation of the birth of Jesus: "And how came Jesus into the world? Through God who created him and woman who bore him. Man, where is your part?" (See Document 43.)

Historians have determined that the most accurate version of Truth's speech was printed in the *Anti-Slavery Bugle,* where it appeared shortly after the event. Another more widely read version appeared in 1881 in *History of Woman Suffrage.* That version, edited by Matilda Joslyn Gage, included the famous passage "Ar'n't I a woman?" and was generally more theatrical and poetic. Sojourner Truth was not the only black woman to attend antebellum women's rights conventions, but she was one of very few, and seems to have been the only black woman to speak. As a chronicler of the conventions Gage was making up in dramatic emphasis what she lacked in content on the topic of black women. Gage's version of the speech became widely known in the 1970s and 1980s, another period of

[107] See Carleton Mabee with Susan Mabee Newhouse, *Sojourner Truth: Slave, Prophet, Legend* (New York: New York University Press, 1993); Nell Irvin Painter, *Sojourner Truth: A Life, A Symbol* (New York: Norton, 1996); and Rosalyn Terborg-Penn, *African-American Women in the Struggle for the Vote, 1850–1920* (Bloomington: Indiana University Press, 1998), 13–35.

heightened interest in women's rights and black rights, when relatively few black women were identified with women's rights.[108]

Abby Price's speech at the Worcester convention of 1850 exemplified the new secular discourse that began to frame women's rights in the 1850s.[109] Price lived with her husband in the "Practical Christian Republic" of Hopedale, Massachusetts, a utopian community with its own journal, the *Practical Christian.* Later they moved to Brooklyn, where she became a friend of Walt Whitman. Her 1850 speech used a new term that came to characterize the women's rights convention movement—"co-equality."[110] In the spirit of the Grimkés she linked co-equality to the equal creation of the sexes. "The natural rights of woman are co-equal with those of man. So God created man in his own image . . . male and female, created he them." Yet, by substituting "co-equal" for the more religious term "moral being," Price and others moved the new women's rights movement forward in ways that did not require participants to accept specific articles of religious faith. Garrisonians or perfectionists could attach religious meaning to "co-equality" if they wished, but others could use the term without connotations that intruded on their religious (or nonreligious) affiliations. To succeed, women's rights had to appeal to a wide range of groups. This term helped them do it.

At Worcester, Abby Price explained that co-equality did not mean that the sexes were the same.

> In contending for this co-equality of woman's with man's rights, it is not necessary to argue, either that the sexes are by nature equally and indiscriminately adapted to the same positions and duties, or that they are absolutely equal in physical and intellectual ability; but only that they are absolutely equal in their rights to life, liberty, and the pursuit of happiness. . . .

Thus, by acknowledging differences between the sexes, co-equality provided the same flexibility that "moral being" had given the Grimkés. The term provided for equal rights, but it also recognized the social and historical reality of sexual difference. (See Document 44.)

One acknowledged difference between women and men that women's rights conventions sought to change was the inferior education of

[108]Mabee, *Sojourner Truth,* 67–82.

[109]For the Worcester convention, see John McClymer, ed., *This High and Holy Moment: The First National Woman's Rights Convention, Worcester, 1850* (New York: Harcourt Brace, 1999). See also Nancy Isenberg, *Sex and Citizenship in Antebellum America* (Chapel Hill: University of North Carolina Press, 1998).

[110]This term seems to have evolved from *coeval,* a word used in the 1848 Resolutions.

women. Advocates of women's co-equality had a difficult time asserting women's intellectual equality because great differences in women's and men's education, especially in the middle and upper classes, meant that men commanded greater intellectual resources. Lucretia Mott wrote in a letter addressed to the Salem, Ohio, women's rights convention:

> Rights are not dependent upon equality of mind; nor do we admit inferiority; leaving that question to be settled by future developments, when a fair opportunity shall be given for the equal cultivation of the intellect, and the stronger powers of the mind shall be called into action.[111]

In this connection, women's conventions, like women's clubs after the Civil War, provided opportunities for women to cultivate their intellects as well as to advocate social change.

Price's secular language showed how American public discourse was absorbing perfectionist ideas, such as "the light of the nineteenth century," and "the progressive spirit of the age." What was once seen as God's dispensation was now attributed to human progress. These views not only broadened the appeal of women's rights, they also helped women's rights discourse fit more easily into the language used by men in legislative halls or political parties.

Women's rights conventions in the 1850s often coincided with state constitutional conventions, just as the women's antislavery conventions of 1837, 1838, and 1839 accompanied the annual meetings of the American Anti-Slavery Society. Lawyers and politicians who supported married women's property rights often attended women's rights conventions, as was the case in 1848 when Elisha Foote, a Seneca falls lawyer who had studied law with Judge Cady, attended the Seneca Falls convention.[112]

One fascinating example of the cultivation of women's rights in other social movements was the positive reception given to a women's rights speaker in the Negro Convention Movement in 1848. Black Americans in the northern states participated actively in the convention culture of antebellum public life. The first National Negro Convention was held in Philadelphia in 1830. Garrison attended the 1831 meeting and contributed to the group's reorientation away from emigration and toward strengthening black communities in the North. In 1843, however, the movement shifted from Garrisonian moral suasion to support the

[111] "From Lucretia Mott to the 'Woman's Convention' to Be Held in Salem, Ohio," *The Liberator,* May 17, 1850.

[112] See Basch, *In the Eyes of the Law.*

Liberty party. The 1848 convention declared Frederick Douglass's *North Star* to be its official national newspaper.[113] His influence at the convention that year may account for the presence of the first woman speaker and the first advocacy of women's rights within the movement. (See Document 45.)

The boundaries between the temperance and women's rights movements were especially permeable because women's auxiliaries within the temperance movement already formed a semi-autonomous movement. A year after the Seneca Falls meeting, *The Lily,* a leading women's temperance periodical, offered a column on "Woman's Rights" that demonstrated the topic's compatibility with temperance. (See Document 46.) When the temperance movement rejected women as delegates to "Men's State Conventions" in 1852, and forced women to take their civic interests elsewhere, many attended women's rights conventions, which often advocated antiliquor legislation. Elizabeth Stanton encouraged connections between the temperance and women's rights movements in the early 1850s by playing a prominent role in women's temperance conventions, frequently chairing the meetings.[114]

The effort of women's rights leaders to recruit other women to their ranks was especially evident in their acknowledgment of the moral reform movement. In their writings and letters, Sarah and Angelina Grimké often referred to "moral reform" as another "great moral enterprise of the day" in addition to antislavery.[115] The Seneca Falls Declaration of Sentiments and resolutions each contained one tribute to the moral reform movement. Resolution six declared "That the same amount of virtue, delicacy, and refinement of behavior, that is required of woman in the social state, should also be required of man, and the same transgressions should be visited with equal severity on both man and woman." The Declaration of Sentiments included a similar passage: "He has created a false public sentiment by giving to the world a different code of morals for men and women." (See Document 42.)

Married women's property rights, temperance, and moral reform were connected to large changes in mid-nineteenth-century America that were redefining middle-class family life and the relationships be-

[113]Howard Holman Bell, *A Survey of the Negro Convention Movement, 1830–1861* (Evanston, Ill.: Howard Holman Bell, 1953; reprint, New York: Arno, 1969), 31–34, 72, 98.

[114]Stanton, *History of Woman Suffrage,* 1: 472–512. See also Dexter C. Bloomer, ed., *Life and Writings of Amelia Bloomer* (Boston: Arena, 1895).

[115]Sarah Grimké, "Dress of Women," in Bartlett, ed., *Letters on the Equality of the Sexes,* 69–70.

tween husbands and wives, parents and children. This connection made women's rights relevant to the daily lives of most middle-class women.

Fundamental to the changes reshaping family life was a long-term decline in birth rates between 1830 and 1940. This decline bridged the transition between traditional and modern patterns of human reproduction. Called the "demographic transition from high birth and death rates to low birth and death rates," it has been studied relatively little by historians of women, even though the transition changed the lives of women profoundly. This decline meant that, on the average, white, native-born women gave birth to half as many children in 1900 as they did in 1800 — 3.5 instead of 7. Through this transition, therefore, women gave less of their own lives to the life of the species. Giving birth less often, they also reduced their risk of dying in childbirth. Eventually affecting all races and all classes, and all modernizing societies, this drop was especially dramatic in the United States because birth rates were higher than other nations in 1800 and fell to be lower than most others by 1940. Two-thirds of this decrease was accomplished by 1880, before the widespread use of contraceptives; the chief methods were sexual abstinence or *coitus interruptus.* Since the decades of most rapid decline in the United States occurred during the 1840s and 1850s, when the society was still overwhelmingly rural, the decline was not caused by industrialization or urbanization.[116]

This remarkable event in human history was achieved by individuals making decisions about their own lives. Culturally speaking, Victorian sexual attitudes helped by encouraging less sexual activity than had been the case before 1800. A good example of those attitudes was Sylvester Graham's 1839 book, *Letters to Young Men,* which, along with dietary recommendations, explained that the expenditure of sperm weakened the male body. He advised limiting the frequency of intercourse to twelve times a year.[117] Elizabeth Stanton encountered the "Graham system" on her honeymoon. She wrote the Grimké sisters that Henry, on their voyage to England, attributed "his freedom from seasickness to *his strict observance of the Graham system*" (her emphasis). More than Henry's diet was probably involved here. (See Document 40.)

[116] See Daniel Scott Smith, "Family Limitation, Sexual Control, and Domestic Feminism in Victorian America," in *Controlling Reproduction: An American History,* ed. Andrea Tone (Wilmington, Del.: SR Books, 1997); and Janet Farrell Brodie, *Contraception and Abortion in 19th-Century America* (Ithaca: Cornell University Press, 1994).

[117] Stephen Nissenbaum, *Sex, Diet, and Debility in Jacksonian America: Sylvester Graham and Health Reform* (Westport, Conn.: Greenwood Press, 1980).

The "Graham system" and other methods of birth control often produced lengthy intervals between births — the most visible evidence being birth-rate declines in the lives of individual women. Elizabeth Stanton had more children than most of her contemporaries (her eighth and last child was born in 1859) but like other women of her generation, she created one extraordinarily long birth interval. After giving birth to three boys at traditional intervals of approximately two years, she maintained a five-and-a-half-year interval between 1845 and 1851.[118] During those years she located herself in a meaningful reform career. Like Harriet Beecher Stowe, who self-consciously maintained a birth interval of six years between 1843 and 1849, during which she established herself as a writer, Stanton resumed bearing children after 1851, integrating them into a life in which they were not the only symbols of her creativity.[119]

Not all men were as modern as Henry Stanton, however. In many marriages women took the initiative to reduce the frequency of intercourse and thereby increase the intervals between births. Women's ability to control their own bodies and to say no to their husbands' sexual demands marked the dawn of a new era. Victorian sexual ideology supported women's control of their bodies, but that ideology was not established without a struggle. In the vanguard of the struggle that promoted the new sexual ideology were members of the American Female Moral Reform Society, which had a branch in almost every city, town, and village in New England and New York in the 1840s.[120] Their campaign against prostitution was also a campaign against predatory male sexuality. (See Document 47.)

By the 1850s these changing sexual mores sustained the careers of an array of professionals. The Reverend Henry Clarke Wright, the Grimkés' advance man in Massachusetts, was one of these. (See Figure 14.) Leaving the abolitionist movement in 1849, he joined the Hopedale Community and began a new career lecturing on "marital abuse"

[118]Historians have looked for but not found evidence to support the possibility that Stanton had miscarriages that would help account for this interval. See Judith Wellman, "The Seneca Falls Women's Rights Convention: A Study of Social Networks," *Journal of Women's History* 3, no. 1 (Spring 1991): 31, note 10.

[119]See Kathryn Kish Sklar, "Victorian Women and Domestic Life: Mary Todd Lincoln, Elizabeth Cady Stanton, and Harriet Beecher Stowe," in Sklar and Dublin, *Women and Power,* 229–43.

[120]See Smith-Rosenberg, "Beauty, the Beast, and the Militant Woman"; Barbara J. Berg, *The Remembered Gate: Origins of American Feminism: The Woman and the City, 1800–1860* (New York: Oxford University Press, 1978); and Wright, "What Was the Appeal of Moral Reform?"

Figure 14 Henry Clarke Wright c. 1858, from the frontispiece of his book, *Marriage and Parentage.* Wright's career in marital reform connected with one of the most fundamental changes in American life — the ability of women to control their own reproductive lives.

and other topics related to the new family order that granted women greater control of their bodies and their lives. His lectures and writings brought a women's rights perspective to marital sexuality. In his 1858 book, *Marriage and Parentage; or, the Reproductive Element in Man, as a Means to His Elevation and Happiness,* he aggressively defended married women's rights to control their own bodies. (See Document 48.) In 1838 Sarah Grimké had touched on this issue when she referred to the "vast amount of secret suffering endured, from the forced submission of women to the opinions and whims of their husbands." (See Document 33.) At the national Negro convention in 1848, Mrs. Sanford spoke of pre-Christian women being "the slave of power and passion."

(See Document 45.) Wright continued these themes in other books—*The Unwelcome Child* (1858) and *The Empire of the Mother over the Character and Destiny of the Race* (1863).[121] His career in the 1850s embodied the remarkable changes that decade wrought in the personal lives of middle-class women, white and black. These widespread changes in women's attitudes toward their bodies established a personal foundation in their lives that led many to support other forms of women's rights.

Ironically, another source of the success of the women's rights convention movement in the 1850s was its less emphatic insistence on racial equality. Even though the compelling importance of racial prejudice had shaped the Grimkés' radicalism and fueled their refusal to retreat on women's rights, many participants in the new movement did not consider racial prejudice a major priority. A resolution on the rights of black women was noticeably absent from the 1848 convention, for example. Although Douglass praised the Seneca Falls Convention in *The North Star,* he did not comment on the absence of women of his own race.

Yet despite the poor showing on race at the Seneca Falls meeting, many leading advocates of women's rights, including Abby Kelley and Frederick Douglass, regularly put the issue of black women's rights before women's rights conventions in the form of resolutions. The Rochester convention of 1848 explicitly called for women's equality without regard to "complexion." And a resolution adopted at Worcester in 1850 read:

> Resolved, That the cause we are met to advocate,—the claim for woman of all her natural and civil rights,—bids us remember the million and a half of slave women at the South, the most grossly wronged and foully outraged of all women; and in every effort for an improvement in our civilization, we will bear in our heart of hearts the memory of the trampled womanhood of the plantation, and omit no effort to raise it to a share in the rights we claim for ourselves.[122]

Abby Kelley, Lucretia Mott, and Amy Post continued to support a dual agenda through the U.S. Civil War, and kept the growing women's rights movement from becoming a white movement. Yet the level of commitment to racial equality was clearly lower in the emerging women's rights movement than it had been among Garrisonians in the 1830s.

[121] Lewis Perry, *Childhood, Marriage, and Reform: Henry Clarke Wright, 1797–1870* (Chicago: University of Chicago Press, 1980), 51–53.

[122] *Proceedings,* Worcester Women's Rights Convention, quoted in McClymer, *This High and Holy Moment,* 132.

No women's rights conventions were held during the conflagration of Civil War between 1861 and 1865. In 1863 Elizabeth Cady Stanton and Susan B. Anthony organized the Women's Loyal National League, which presented a petition with 300,000 signatures to the U.S. Senate demanding the abolition of slavery by a constitutional amendment. In 1866 at the first women's rights convention after the war, women's rights activists formed the American Equal Rights Association (AERA) to continue the antebellum solidarity between women's rights and abolition and to promote universal suffrage — suffrage for black men as well as all women.

At the 1867 meeting of the AERA, Lucretia Mott, its president, looked back on twenty years of women's growing public power. "In the temperance reformation, and in the great reformatory movements of our age, woman's powers have been called into action," she said. "They are beginning to see that another state of things is possible for them, and they are beginning to demand their rights." Noting that they were meeting in one of the wealthiest and most prestigious churches in America — the Church of the Pilgrims in Brooklyn Heights — Mott asked, "Why should this church be granted for such a meeting as this, but for the progress of the cause? Why are so many present, ready to respond to the most ultra and most radical sentiments here, but that woman has grown and is able to assume her rights?" What had been "ultra" and "radical" in the 1830s had almost become respectable by 1867.[123]

Women's rights emerged in the 1830s in a religious context in which women speakers were challenging themselves and their audiences to act on their professed beliefs about human equality — the equality of white and black Americans, both slave and free, and the equality of men and women. Out of that religious crucible new ideas were born. Those ideas found fertile soil in the profound changes taking place in American society at midcentury, especially in the evolution of American civic life and in the reconstruction of family life.

Through the convention movement, women's rights took root in U.S. political culture. The movement thereafter drew upon both the assets and limitations of American public life. Its struggle to extend suffrage rights to white women championed one of the fundamental features of American life — self-governance. Yet the tenuousness with which the

[123] Lucretia Mott quoted in *History of Woman Suffrage,* 2: 199. This church was the chief competitor to Henry Ward Beecher's Plymouth Church. The average personal property value of members of the Church of the Pilgrims ($200,000) was four times that of Beecher's parishioners. Altina L. Waller, *Reverend Beecher and Mrs. Tilton: Sex and Class in Victorian America* (Amherst: University of Massachusetts Press, 1982), 105.

new movement maintained the historic alliance that the Garrisonians had forged with free blacks revealed their difficulty in rising above the characteristics of the political culture that they were joining.

Women's rights arose within a relatively small group of women and men who expressed a moment of high idealism in American public life. Aspects of that idealism persisted in the movement. What began in the 1830s moved forward into the 1870s as one of the most vital characteristics of American society: the mobilization of women — black and white — within organizations that shaped the public life of their communities and their nation.

EPILOGUE: THE NEW MOVEMENT SPLITS OVER THE QUESTION OF RACE, 1850–1869

The Garrisonian coalition between black rights and women's rights encountered difficulties almost immediately in the women's rights convention movement of the 1850s. One leading abolitionist woman publicly objected to this coalition. Jane Swisshelm (1815–84), for ten years the editor of Pittsburgh's only abolitionist newspaper, *The Saturday Visiter,* was a strong supporter of women's rights; her editorials in support of married women's property rights contributed to the enactment of such rights in Pennsylvania in 1848. Yet in a column about the Worcester Convention she strenuously denounced the resolution on behalf of slave women, writing that "In a Woman's Rights Convention the question of color had no right to a hearing." (See Document 49.) Parker Pillsbury, a Garrisonian friend of Abby Kelley and Frederick Douglass, criticized Swisshelm's column in a letter to *The North Star.*[124] He admonished Swisshelm to remember that unless black women were specifically mentioned, they would not be thought of at the convention, and reminded her that "scarcely a colored person, man or woman, appeared" in the convention. (See Document 50.) She replied that the issues of race and sex should be separated, arguing that the social status of women was set by the men of their class and race. (See Document 51.)

While the possibility of a cross-race coalition was debated in the women's rights movement in the 1850s, during the Civil War this and other questions were suspended as participants in the movement turned

[124] For Pillsbury, see Stacey Marie Robertson, "Parker Pillsbury, Antislavery Apostle: Gender and Religion in Nineteenth-Century United States Radicalism," Ph.D. diss. (Santa Barbara: University of California Press, 1994); for Swisshelm, see *Notable American Women.*

their energies to supporting the Union cause. However, in a series of meetings after the war, the movement split over the question of race. In fact, two splits occurred. One was an informal division into a white movement and a black movement, which occurred gradually between 1867 and 1890. Another was the formal division between those who did and did not support the passage of the Fifteenth Amendment to the U.S. Constitution, which in 1870 prohibited states from denying citizens the right to vote "on account of race, color, or previous condition of servitude." Because the amendment did not mention sex, and therefore was not interpreted as applying to women, Stanton, Anthony, and their supporters opposed its passage.

Racial differences and animosities were evident in New York City in 1866, at the first post-war women's rights convention. There Frances Ellen Watkins Harper (1825–1911) was the most prominent black woman speaker. (See Figure 15.) A free black from Baltimore, Harper had moved to Philadelphia in 1851 to avoid the effects of the Fugitive Slave Law, passed in 1850, which made it easier for free blacks to be kidnapped into slavery. She became an antislavery lecturer in 1854, and, through her essays and poetry, a leading African American writer. Garrison wrote an introduction to her *Poems on Miscellaneous Subjects.*[125]

Harper recognized the need for black women's rights, but also expressed her irritation with white women. "If there is any class of people who need to be lifted out of their airy nothings and selfishness, it is the white women of America," she said. (See Document 52.) Speaking for the needs of black men, Harper referred to the Dred Scott decision of the U.S. Supreme Court, which in 1857 had found that "men of my race had no rights which the white man was bound to respect." This decision demonstrated to Harper the urgent need for federal protection of the rights of black men. The Fourteenth Amendment to the U.S. Constitution, then being debated in Congress and ultimately ratified in 1868, would be the means by which the Dred Scott decision was overcome and the rights of black men guaranteed. It would also be the means by which the word "male" was first embedded into the Constitution.[126]

Participants in that 1866 convention overcame their differences to create the American Equal Rights Association (AERA), dedicated to universal suffrage and to the support of both black and woman suffrage. Yet

[125] Frances Ellen Watkins, *Poems on Miscellaneous Subjects* (Boston: Yerrinton, 1854).

[126] For the context of Harper's speech, see Rosalyn Terborg-Penn, *African American Women in the Struggle for the Vote, 1850–1920* (Bloomington: Indiana University Press, 1998), 13–35.

Figure 15 Frances Ellen Watkins Harper, a leading abolitionist lecturer to black and racially mixed audiences on the eve of the Civil War, resumed lecturing after the war, emphasizing the need for education, temperance, and family renewal among former slaves. She was outspoken in denouncing white racism and violence.
Boston Athenæum.

at the 1869 meeting of the AERA, the cross-race coalition foundered. Debate there produced a logjam of disagreement rather than a basis for future cooperation. Competing referendums on woman suffrage and black suffrage in Kansas in 1867 had set those two constituencies against one another. The ratification of the Fourteenth Amendment in 1868, which guaranteed citizenship rights to "male" residents of states, deepened the antagonism. In February 1869, Congress had passed the Fifteenth Amendment, which explicitly enfranchised black men, and that spring the amendment was in the process of being ratified by the states. Stanton and Anthony opposed the amendment; Douglass, Harper, and Stone supported it. Speakers in this debate probably realized that their disagreements marked the demise of the AERA.[127] (See Document 53.)

This debate was largely framed in terms of the opposing claims of "the woman" and "the black man." For Frederick Douglass the needs of black women were absorbed into those of black people. Black women were oppressed, he argued, not because they were women, but because they were black. Lucy Stone added that one great ocean of wrong engulfed "the black man," the other "the woman." Paulina Davis, who organized the Worcester women's rights convention in 1850, called for a Sixteenth Amendment that would protect black women in the South against "a race of tyrants raised above her." Elizabeth Cady Stanton argued against enfranchising "ignorant negroes and foreigners" until women were admitted to the polls. Frances Harper commented that "the white women all go for sex, letting race occupy a minor position." Susan Anthony opposed the Fifteenth Amendment because it would put black men "in position of tyrants over" black women "who had until now been the equals of the men at their side." (See Document 53.)

In 1881 the editors of *History of Woman Suffrage* concluded that these debates "proved the futility of any attempt to discuss the wrongs of different classes in one association." (See Document 54.) Although this statement meant to point to the differences between "black men" and "women," it also suggested that differences between black women and white women were too large to be encompassed by one organization.

In 1869 two woman suffrage organizations were created. The National Woman Suffrage Association, located in New York and headed by Elizabeth Cady Stanton and Susan B. Anthony, opposed the Fifteenth Amendment. The American Woman Suffrage Association, headquartered in Boston and led by Elizabeth Blackwell and Lucy Stone, supported the amendment. The legacy of this difference on the question of black male suffrage was not overcome until 1890, when these two groups finally united to form the National American Woman Suffrage Association.

[127] For this split, see Ellen Carol DuBois, ***Feminism and Suffrage: The Emergence of an Independent Women's Movement in America, 1848–1869*** (Ithaca: Cornell University Press, 1978), 162–202; and Andrea Moore Kerr, ***Lucy Stone: Speaking Out for Equality*** (New Brunswick, N.J.: Rutgers University Press, 1992), 138–59.

After 1869, Sojourner Truth affiliated with Elizabeth Stanton and the National Woman Suffrage Association, and Frances Harper associated with Lucy Stone and the American Woman Suffrage Association.[128] But Harper, Truth, and other black women remained marginal within white woman suffrage organizations. After 1870 other black women, including Ida B. Wells, formed separate black women's suffrage associations.[129]

Just as the "woman question" split the antislavery movement in the late 1830s, so the race question split the women's rights movement in the late 1860s. These fractures revealed important aspects of American civil society. But just as significant were the forces that forged coalitions across the differences of race and gender. These too disclosed notable features of American public life.

[128] Carla L. Peterson, *"Doers of the Word": African-American Women Speakers and Writers in the North (1830–1880)* (New York: Oxford University Press, 1995), 120, 199, 224, 228–29; Bettye Collier-Thomas, "Frances Ellen Watkins Harper: Abolitionist and Feminist Reformer, 1825–1911," in *African American Women and the Vote, 1837–1965,* ed. Ann D. Gordon et al. (Amherst: University of Massachusetts Press, 1997), 41–65; and Frances Smith Foster, ed., *A Brighter Coming Day: A Frances Ellen Watkins Harper Reader* (New York: Feminist Press, 1990).

[129] See Terborg-Penn, *African American Women,* 36–54.

PART TWO

The Documents

SEEKING A VOICE: GARRISONIAN ABOLITIONIST WOMEN, 1831–1833

1

LUCRETIA MOTT

Life and Letters

1884

Lucretia Mott, a Quaker minister and abolitionist leader, years later vividly recalled her participation in the founding meeting of two important Garrisonian abolitionist groups in 1833: the American Anti-Slavery Society and the Philadelphia Female Anti-Slavery Society. While she had considerable experience in Quaker groups before 1833, including groups of Quaker women, meeting with non-Quakers in these antislavery societies was a new and somewhat strange experience.

Although we were not recognized as a part of the convention by signing the document, yet every courtesy was shown to us, every encouragement given to speak, or to make suggestions of alteration. I do not think it occurred to any one of us at that time, that there would be a propriety

---

Anna Davis Hallowell, *James and Lucretia Mott, Life and Letters* (Boston: Houghton Mifflin, 1884), 114, 121.

in our signing the document. It was with difficulty, I acknowledge, that I ventured to express what had been near to my heart for many years, for I knew we were there by sufferance; but when I rose, such was the readiness with which the freedom to speak was granted, that it inspired me with a little more boldness to speak on other subjects. When the declaration was under consideration, and we were considering our principles and our intended measures of action, when our friends felt that they were planting themselves on the truths of Divine Revelation, and on the Declaration of Independence, as an Everlasting Rock, it seemed to me, as I heard it read, that the climax would be better to transpose the sentence and place the Declaration of Independence first, and the truths of Divine Revelation last, as the Everlasting Rock; and I proposed it. I remember one of the younger members turning to see what woman there was there who knew what the word "transpose" meant. . . .

At that time I had no idea of the meaning of preambles, and resolutions, and votings. Women had never been in any assemblies of the kind. I had attended only one convention — a convention of colored people — before that; and that was the first time in my life I had ever heard a vote taken, being accustomed to our Quaker way of getting the prevailing sentiment of the meeting.[1] When, a short time after, we came together to form the Female Anti-Slavery Society, there was not a woman capable of taking the chair and organizing that meeting in due order; and we had to call on James McCrummel, a colored man, to give us aid in the work.

[1] In Quaker meetings decisions were usually made by consensus.

2

Constitution of the Afric-American Female Intelligence Society

1831

Free black women were among the first Americans to organize antislavery organizations devoted to immediate emancipation. Although this state-

The Liberator, Jan. 7, 1832.

ment of purpose emphasized mutual assistance, they were a Garrisonian group.

Whereas the subscribers, women of color of the Commonwealth of Massachusetts, actuated by a natural feeling for the welfare of our friends, have thought fit to associate for the diffusion of knowledge, the suppression of vice and immorality, and for cherishing such virtues as will render us happy and useful to society, sensible of the gross ignorances under which we have too long labored, but trusting, by the blessing of God, we shall be able to accomplish the object of our union—we have therefore associated ourselves under the name of the Afric-American Female Intelligence Society.

3

MARIA STEWART

Religion and the Pure Principles of Morality

1831

Ardently asserting the rights of black people to participate fully in the opportunities of American life, in this pamphlet Maria Stewart, an African American writer and lay preacher, urged black women to unite economically and politically.

I was born in Hartford, Connecticut, in 1803; was left an orphan at five years of age; was bound out in a clergyman's family;[1] had the seeds of piety and virtue early sown in my mind, but was deprived of the advantages of education, though my soul thirsted for knowledge. Left them at

[1] "Bound out" refers to the process by which local authorities placed orphans in families who agreed to provide room, board, and clothing in exchange for relief payments and the child's labor.

Pamphlet published by *The Liberator*, Oct. 8, 1831. Reprinted in Maria Stewart, *Meditations from the Pen of Mrs. Maria W. Stewart* (1879), and in Marilyn Richardson, ed., *Maria W. Stewart: America's First Black Woman Political Writer: Essays and Speeches* (Bloomington: University of Indiana Press, 1987), 28–42.

fifteen years of age; attended Sabbath schools until I was twenty; in 1826 was married to James W. Stewart; was left a widow in 1829; was, as I humbly hope and trust, brought to the knowledge of the truth, as it is in Jesus, in 1830; in 1831 made a public profession of my faith in Christ.

From the moment I experienced the change, I felt a strong desire, with the help and assistance of God, to devote the remainder of my days to piety and virtue, and now possess that spirit of independence that, were I called upon, I would willingly sacrifice my life for the cause of God and my brethern. . . .

All the nations of the earth are crying out for liberty and equality. Away, away with tyranny and oppression! And shall Afric's sons be silent any longer? Far be it from me to recommend to you either to kill, burn, or destroy. But I would strongly recommend to you to improve your talents; let not one lie buried in the earth. Show forth your powers of mind. . . .

This is the land of freedom. The press is at liberty. Every man has a right to express his opinion. Many think, because your skins are tinged with a sable hue, that you are an inferior race of beings; but God does not consider you as such. He hath formed and fashioned you in his own glorious image, and hath bestowed upon you reason and strong powers of intellect . . . and according to the Constitution of these United States, he hath made all men free and equal. Then why should one worm say to another, "Keep you down there, while I sit up yonder; for I am better than thou?" It is not the color of the skin that makes the man, but it is the principles formed within the soul. . . .

O, ye daughters of Africa, awake! Awake! Arise! No longer sleep nor slumber, but distinguish yourselves. Show forth to the world that ye are endowed with noble and exalted faculties. . . . Let every female heart become united, and let us raise a fund ourselves; and at the end of one year and a half, we might be able to lay the corner stone for the building of a High School, that the higher branches of knowledge might be enjoyed by us. . . . How long shall the fair daughters of Africa be compelled to bury their minds and talents beneath a load of iron pots and kettles? Until unity, knowledge and love begin to flow among us. . . .

The Americans have practised nothing but head-work these 200 years, and we have done their drudgery. And is it not high time for us to imitate their examples, and practice head-work too? . . .

It is the blood of our fathers, and the tears of our brethren that have enriched your soils. AND WE CLAIM OUR RIGHTS. We will tell you that we are not afraid of them that kill the body, and after that can do no more.

4

MARIA STEWART

Lecture Delivered at the Franklin Hall

Boston, 1832

Stewart's address at Franklin Hall challenged black and white women to consider their social status in new ways, and to find new insights and understandings about the effects of racial prejudice. The geographic location of Franklin Hall suggests that Stewart's talk was not strongly supported by Boston's African American churches, since the hall stood in South Boston, not in the West End of the city where the highest concentration of African Americans lived and where most African American churches were located.[1]

Methinks I heard a spiritual interrogation—"Who shall go forward, and take off the reproach that is cast upon the people of color? Shall it be a woman?" And my heart made this reply—"If it is thy will, be it even so, Lord Jesus!" . . .

I have asked several individuals of my sex, who transact business for themselves, if providing our girls were to give them the most satisfactory references, they would not be willing to grant them an equal opportunity with others? Their reply has been—for their own part, they had no objection; but as it was not the custom, were they to take them into their employ, they would be in danger of losing the public patronage.

And such is the powerful force of prejudice. Let our girls possess whatever amiable qualities of soul they may; let their characters be fair and spotless as innocence itself; let their natural taste and ingenuity be what they may; it is impossible for scarce an individual of them to rise above the condition of servants. Ah! why is this cruel and unfeeling distinction? Is it merely because God has made our complexion to vary? If it be, O shame to soft relenting humanity! . . . Yet, after all, methinks

[1]See Adelaide M. Cromwell, "The Black Presence in the West End of Boston, 1800–1864: A Demographic Map," in Donald M. Jacobs, ed., *Courage and Conscience: Black and White Abolitionists in Boston* (Bloomington: Indiana University Press, 1993), pp. 164–65; and John Toomey and Edward Rankin, *History of South Boston* (Boston, 1901).

Reprinted in Richardson, *Maria W. Stewart,* 45–48.

were the American free people of color to turn their attention more assiduously to moral worth and intellectual improvement, this would be the result: prejudice would gradually diminish, and the whites would be compelled to say, unloose those fetters! . . .

O, ye fairer sisters, whose hands are never soiled, whose nerves and muscles are never strained, go learn by experience! Had we had the opportunity that you had had, to improve our moral and mental faculties, what would have hindered our intellects from being as bright, and our manners from being as dignified as yours? Had it been our lot to have been nursed in the lap of affluence and ease, should we not have naturally supposed that we were never made to toil? And why are not our forms as delicate, and our constitutions as slender, as yours? Is not the workmanship as curious and complete? Have pity upon us, have pity upon us, O ye who have hearts to feel for other's woes; for the hand of God has touched us.

5

MARIA STEWART

Farewell Address to Her Friends in the City of Boston

1833

As she prepared to leave Boston, Stewart's farewell address expressed her belief that her leadership had been rejected by the black community because she was a woman. Yet she also presented a spirited defense of her female identity and her ability to advance the interests of free black people.

On my arrival here, not finding scarce an individual who felt interested in these subjects [the welfare of the black community], and but few of the whites, except Mr. Garrison, and his friend, Mr. Knapp; and hearing that those gentlemen had observed that female influence was powerful; my soul became fired with a holy zeal for your cause. . . . The spirit of God came before me, and I spake before many. When going home,

The Liberator, Sept. 28, 1833; reprinted in Richardson, *Maria W. Stewart*, 65–74.

reflecting on what I had said, I felt ashamed, and knew not where I should hide myself. A something said within my breast, "Press forward, I will be with thee." And my heart made this reply, "Lord, if thou wilt be with me, then I will speak for thee as long as I live." . . .

What if I am a woman; is not the God of ancient times the God of these modern days? Did he not raise up Deborah, to be a mother, and a judge in Israel? Did not queen Esther save the lives of the Jews? And Mary Magdalene first declare the resurrection of Christ from the dead? . . . St. Paul declared that it was a shame for a woman to speak in public. . . . Did St. Paul but know of our wrongs and deprivations, I presume he would make no objections to our pleading in public for our rights. . . .

If such women as are here described have once existed, be no longer astonished then, my brethren and friends, that God at this eventful period should raise up your own females to strive, by their example both in public and private, to assist those who are endeavoring to stop the strong current of prejudice that flows so profusely against us at present. No longer ridicule their efforts, it will be counted for sin. For God makes use of feeble means sometimes, to bring about his most exalted purposes. . . .

Dearly beloved, I have made myself contemptible in the eyes of many, that I might win some. But it has been like labor in vain. . . . The bitterness of my soul has departed from those who endeavored to discourage and hinder me in my Christian progress; and I can now forgive my enemies, bless those who have hated me, and cheerfully pray for those who have dispitefully used and persecuted me.

FARE YOU WELL, FAREWELL.
MARIA STEWART

WOMEN CLAIM THE RIGHT TO ACT: ANGELINA AND SARAH GRIMKÉ SPEAK IN NEW YORK, JULY 1836–MAY 1837

6

AMERICAN ANTI-SLAVERY SOCIETY

Petition Form for Women

1834

By 1834 women had become so active in the petition campaign to Congress to end slavery in the District of Columbia that the AASS printed a special form for them. While this form did not explicitly assert women's right to petition, it did argue that women's petitions were appropriate to the political moment. Hundreds of women who signed these forms thronged to hear the Grimké sisters speak in 1836 and 1837.

FATHERS AND RULERS OF OUR COUNTRY:
Suffer us, we pray you, with the sympathies which we are constrained to feel as wives, as mothers, and as daughters, to plead with you in behalf of a long oppressed and deeply injured class of native Americans [i.e., American-born slaves], residing in that portion of our country which is under your exclusive control. We should poorly estimate the virtues which ought ever to distinguish your honorable body could we anticipate any other than a favorable hearing when our appeal is to men, to philanthropists, to patriots, to the legislators and guardians of a Christian people. We should be less than women, if the nameless and unnumbered wrongs of which the slaves of our sex are made the defenseless victims, did not fill us with horror and constrain us, in earnestness and agony of spirit to pray for their deliverance. By day and by night, their woes and wrongs rise up before us, throwing shades of mournful contrast over the joys of domestic life, and filling our hearts with sadness at the recollection of those whose hearths are desolate.

Gilbert H. Barnes and Dwight L. Dumond, ed., *Letters of Theodore Dwight Weld, Angelina Grimké Weld and Sarah Grimké, 1822–1844* (New York: Appleton-Century-Crofts, 1934; reprint Gloucester, Mass.: Smith, 1965), 1:175–76. [Hereafter *Weld-Grimké Letters.*]

Nor do we forget, in the contemplation of their other sufferings, the intellectual and moral degradation to which they are doomed; how the soul formed for companionship with angels, is despoiled and brutified, and consigned to ignorance, pollution, and ruin.

Surely then, as the representatives of a people professedly christian, you will bear with us when we express our solemn apprehensions in the language of the patriotic Jefferson "we tremble for our country when we remember that God is just, and that his justice cannot sleep forever," and when in obedience to a divine command "we remember them who are in bonds as bound with them." Impelled by these sentiments, we solemnly purpose, the grace of God assisting, to importune high Heaven with prayer, and our national Legislature with appeals, until this christian people abjure forever a traffic in the souls of men, and the groans of the oppressed no longer ascend to God from the dust where they now welter.

We do not ask your honorable body to transcend your constitutional powers, by legislating on the subject of slavery within the boundaries of any slaveholding State; but we do conjure you to abolish slavery in the District of Columbia where you exercise exclusive jurisdiction. In the name of humanity, justice, equal rights and impartial law, our country's weal, her honor and her cherished hopes we earnestly implore for this our humble petition, your favorable regard. If both in christian and in heathen lands, Kings have revoked their edicts, at the intercession of woman, and tyrants have relented when she appeared a suppliant for mercy, surely we may hope that the Legislators of a free, enlightened and christian people will lend their ear to our appeals, when the only boon we crave is the restoration of rights unjustly wrested from the innocent and defenseless.—And as in duty bound your petitioners will ever pray.

7

ANGELINA GRIMKÉ

Appeal to the Christian Women of the South

1836

In her first publication, Angelina Grimké expressed themes that characterized women's abolitionist efforts, especially their appeal to women as women and their encouragement of women's activism in public life.

RESPECTED FRIENDS:
It is because I feel a deep and tender interest in your present and eternal welfare that I am willing thus publicly to address you. Some of you have loved me as a relative, and some have felt bound to me in Christian sympathy, and Gospel friendship. . . . It is because you have known me, that I write thus unto you. . . .

If you really suppose *you* can do nothing to overthrow slavery, you are greatly mistaken. You can do much in every way: four things I will name. 1st. You can read on this subject. 2d. You can pray over this subject. 3d. You can speak on this subject. 4th. You can *act* on this subject. . . .

3. Speak on this subject. It is through the tongue, the pen, and the press, that truth is principally propagated. Speak then to your relatives, your friends, your acquaintances on the subject of slavery; be not afraid if you are conscientiously convinced it is sinful, to say so openly, but calmly, and let your sentiments be known. If you are served by the slaves of others, try to ameliorate their condition as much as possible; never aggravate their faults, and thus add fuel to the fire of anger already kindled in a master and mistress's bosom. . . . Discountenance *all* cruelty to them, all starvation, all corporal chastisement; these may brutalize and *break* their spirits, but will never bond them to willing, cheerful obedience. If possible, see that they are comfortably and *seasonably* fed, whether in the house or the field; it is unreasonable and cruel to expect slaves to wait for their breakfast until eleven o'clock, when they rise at five or six. Do all you can, to induce their owners to clothe them well, and

---

Angelina E. Grimké, *Appeal to the Christian Women of the South* (New York: [American Anti-Slavery Society], 1836; reprint New York: Arno, 1969).

to allow them many little indulgences which would contribute to their comfort. Above all, try to persuade your husband, father, brothers and sons, that *slavery is a crime against God and man* . . . be faithful in pleading the cause of the oppressed.

> Will you behold unheeding,
> Life's holiest feeling crushed,
> Where *woman's* heart is bleeding,
> Shall *woman's* heart be hushed? . . .

4. Act on this subject. Some of you *own* slaves yourselves. If you believe slavery is *Sinful,* set them at liberty, "undo the heavy burdens and let the oppressed go free." If they wish to remain with you, pay them wages, if not let them leave you. Should they remain teach them and have them taught the common branches of an English education; they have minds and those minds *ought to be improved.* So precious a talent as intellect, never was given to be wrapt in a napkin and buried in the earth. It is the *duty* of all, as far as they can, to improve their own mental faculties, because we are commanded to love God with *all our minds,* as well as with all our hearts, and we commit a great sin, if we *forbid or prevent* that cultivation of the mind in others, which would enable them to perform this duty. . . .

I know that this doctrine of obeying *God,* rather than man, will be considered as dangerous and heretical by many, but I am not afraid openly to avow it, because it is the doctrine of the Bible. . . . If a law commands me to sin I will break it. . . . The doctrine of blind obedience and unqualified submission to *any human* power, whether civil or ecclesiastical, is the doctrine of despotism, and ought to have no place among Republicans and Christians.

But you will perhaps say, such a course of conduct would inevitably expose us to great suffering. Yes! my christian friends, I believe it would, but this will *not* excuse you or anyone else for the neglect of *duty.* If Prophets and Apostles, Martyrs, and reformers had not been willing to suffer for truth's sake, where would the world have been now? If they had said, we cannot speak the truth, we cannot do what we believe is right, because *the laws of our country or public opinion are against us,* where would our holy religion have been now? . . . Why were the Presbyterians chased like the partridge over the highlands of Scotland — the Methodists pumped, and stoned, and pelted with rotten eggs — the Quakers incarcerated in filthy prisons, beaten, whipped at the cart's tail, banished and hung? Because they dared to *speak* the *truth,* to *break* the unrighteous *laws* of their country, and chose rather to suffer affliction with the people of God, "not accepting deliverance," even under the gallows.

But you may say we are *women*, how can *our* hearts endure persecution? And why not? Have not women stood up in all the dignity and strength of moral courage to be the leaders of the people, and to bear a faithful testimony for the truth whenever the providence of God has called them to do so? . . .

Let [the Christian women of the South] embody themselves in societies, and send petitions up to their different legislatures, entreating their husbands, fathers, brothers, and sons, to abolish the institution of slavery; no longer to subject *woman* to the scourge and the chain, to mental darkness and moral degradation, no longer to tear husbands from their wives, and children from their parents; no longer to make their lives bitter in hard bondage; no longer to reduce *American citizens* to the abject condition of *slaves*, of "chattels personal"; no longer to barter the *image of God* in human shambles for corruptible things such as silver and gold. . . .

If you could obtain but six signatures to such a petition in only one state, I would say, send up that petition, and be not in the least discouraged by the scoffs and jeers of the heartless, or the resolution of the house to lay it on the table. It will be a great thing if the subject can be introduced into your legislatures in any way, even by *women*, and *they* will be the most likely to introduce it there in the best possible manner, as a matter of *morals* and *religion*, not of expediency or politics. . . .

I have appealed to your sympathies as women, to your sense of duty as *Christian women*. . . . Count me not your "enemy because I have told you the truth," but believe me in unfeigned affection,

YOUR SYMPATHIZING FRIEND,
ANGELINA E. GRIMKÉ

8

ANGELINA GRIMKÉ

Letter to Jane Smith

New York, December 17, 1836

In a series of personal letters to Jane Smith, her best friend, Angelina Grimké expressed her hopes and fears about her public speaking in New York. Smith was a Quaker who joined the Philadelphia Female Anti-Slavery Society in 1838.

MY BELOVED JANE:

Thou deservest a good long letter & I feel far more capable of writing such a one today than I have for some weeks past & now I am going to tell thee all my heart. After the privilege of attending the convention (during which time my feelings were too much occupied with the business & taking notes to have much time to think of myself), I began to feel afresh my *utter inability* to do any thing in the work I had undertaken. The more I looked at it, with the eye of reason, the more unnatural it seemed, & if I had dared to return to Philadelphia & lay down my commission, most gladly would I have done so. But one little grain of faith yet remains. I remember the deep travail of spirit thro' which I had passed at Shrewsbury & could not but believe that He who had sent me out would go before & prepare the way of the poor instruments he was pleased to employ.

Last week a Baptist minister of the name of Dunbar proposed our having a meeting in his Session room. . . . This was a great relief to our minds, for we both felt that this was just the right thing & readily closed in with the offer. The Female A S Sy embraced the opportunity of making this the commencement of Quarterly Meetings for their Sy, & it was accordingly given out in 4 churches on the Sabbath, but our names *not* mentioned. Well, after this was done we felt almost in despair about the meeting, for we know that some persons here were exceedingly afraid that if we addressed our sisters, it would be called Quaker preaching &

Weld-Grimké Papers, Clements Library, University of Michigan, Ann Arbor. [Hereafter Weld-Grimké Papers.]

that the prejudice here against women speaking in public life was so great that if such a view was taken, our precious cause would be injured.

The Throne of Grace was our only refuge & to it we often fled in united supplication for divine help. On 4th day morning, our dear brother in the Lord, G. Smith, came to henry Ludlow's (where we are now staying) to breakfast, & when we gathered round the family altar, our hearts were melted together as he poured out his soul in prayer for *us,* particularly, that we might be directed, strengthened & comforted in our work of mercy, & as soon as we rose from our places, the bell rung & a printed notice of the meeting was handed in, & in it our names were mentioned as intending to address the meeting. It was too much for us, & in christian freedom we opened our hearts to our dear friends. Truly we felt as if such a thing was humanly impossible. We talked the matter over & found that G Smith had another fear, that it would be called a Fanny Wright meeting & so on, & advised us not to make addresses except in parlors.

Well, we did not know what to do. The meeting was appointed & there was no business at all to come before it. When he left us, I went to my room. I laid my difficulty at the feet of Jesus. I called upon him in my trouble & he harkened unto my cry, renewed my strength & confidence in God, & from that time I felt sure of his help in the hour of need. My burden was rolled off upon his everlasting arm, & I could rejoice in a full assurance of his mercy & power to be mouth & wisdom, tongue & utterance to us both.

Yesterday morning, T d Weld came up like a brother to sympathize with us & encourage our hearts in the Lord. He is a precious christian, bid us not to fear, but to trust in God &c. In a previous conversation on our holding meetings, he had expressed his full unity with our doing so, and grieved over that factitious state of society which bound up the energies of woman, instead of allowing her to exercise them to the glory of God and the good of her fellow creatures. In the cause of the slaves, he believes, she has a *great* work to do & *must* be awakened to her responsibility &c.

His visit was really a strength to us, & I felt *no* fear about the consequences, went to the meeting at 3 O clock & found about 300 persons. It was opened with prayer by H Ludlow. We were warmly welcomed by brother Dunbar. They soon left us &, after an opening minute, I spoke for about 40 minutes, I think, feeling perfectly unembarrassed, about which Dear Sister [Sarah] did her part better than I did. We then read some extracts from papers & letters & answered a few questions, when at 5 the meeting closed, after the question had been put whether our sisters wished another meeting to be held. A good many rose & H L says he is sure he can get his Session room for us. . . . Many came up and spoke to us after the meeting was over. . . .

I know nothing of the effects on *others*. We went home with Julia Tappan to tea, & brother Weld was all anxiety to know about it. She undertook to give some account & among other things mentioned that a warm-hearted Abolitionist had found *his* way into the back pack of the meeting & that H L had escorted him out. Weld's countenance was instantly lighted up, & he exclaimed how extremely ridiculous to think of a man's being shouldered out of a meeting for fear he should hear a woman speak. *We* smiled & said we did not know how it seemed to others, but it looked very strange in our eyes. . . .

[Gerrit Smith] is one of the noblest, loveliest men I ever met. He seemed just like a brother to us and invited us very kindly to go to their house as a resting place after our winter work was over. No doubt thou will want to know when we expect to leave N Y. We don't know. Sister is now writing an Address to Southern Clergymen & we think this had best be finished before we go hence. She has already written 24 p & it is not yet done. Then it must all be copied, so that we cannot tell any thing about our movements. . . . We hope that now beginning has been made, that we shall be able to hold a series of meetings here with our sisters, &, as the brethern think it will not do to have public lectures in N Y, it seems the more necessary for us to do what we can. . . .

VERY AFFY A E GÉ

9

ANGELINA GRIMKÉ

Letter to Jane Smith

New York, January 20, 1837

To Jane Smith, Angelina Grimké described her growing power as a public speaker, the deepening effect the sisters were having on their audiences, and the commitment they were developing to their work.

MY DEAR JANE:
For the three weeks previous we had lectured on the Laws of the Slave States & illustrated each by example to show those laws were not a dead letter.[1] Yesterday we had intended to close this part of the subject by this testimony, then by showing that Slavery is cruel to the body, heart, mind and soul of the slave. But I could not get thro' more than the two first, so that the degradation of the mind and destruction of the soul remain for next week. We now hold our meetings regularly at Henry G. Ludlow's session room every 5th day [Thursday] afternoon at 3 O'Clock. By the by, as the room was so crowded and oppressingly warm he gave out that we should have the Church itself hereafter—there must have been *more* than 300 out yesterday & we are told, a more influential class than at first attended. Our publications were eagerly received by the hands which were raised to catch them as we threw them into the crowd. A deepening interest we think is evidently exhibited. It really seems as if the Lord was moving by his Spirit on the hearts of the people and that the tide of feeling is beginning to rise, which under the Divine Blessing may yet move this city to rise up in the dignity of moral power against the crying sin of our Land.

But dear friend, thou will doubtless want to know whether I find it an *easy* thing to hold such meetings—I, no! I can truly say that the day I have to speak is always a day of suffering, & I now understand what friends mean when they say, they speak for *the relief of their own minds.* I feel like a totally different being after the meeting is over, for I assure

[1] Hence AG spoke against the claims of pro-slavery advocates that cruel slave laws were not enforced.

Weld-Grimké Papers.

thee I do know that a fresh caption [source of energy] is needed for every appearance in public. It is really delightful to see dear Sister so happy in this work. I have not the shadow of a doubt she is in her right place & will be made instrumental of great good. . . .

Thou mayest remark I speak of our *talks* as *lectures*. Well this is the name that *others* have given our poor effort, & I don't know in fact what to call such novel proceedings. How little! how *very little* I supposed, when I used to say "I wish I was a man, that I might go out and lecture," that I would ever do such a thing. The idea never crossed my mind that *as a woman* such work could possibly be assigned me. But the Lord is "wonderful counsel, excellent in working," making a way for his people when there seems to be *no* way. Dear Jane, I love the work. I count myself greatly favored in being called to it, & I often feel as if the only earthly blessing I have to ask for is to be made the unworthy instrument of arousing the slumbering energy & dormant sympathy of my northern sisters on this deeply painful & interesting subject. . . .

10

ANGELINA GRIMKÉ

Letter to Jane Smith

New York, February 4, 1837

In this letter Angelina Grimké first mentioned that she was referring to women's rights in her lectures. Speaking in non-Quaker religious institutions, the Grimkés were doubly conscious of how they were deviating from accepted social norms.

MY DEAR JANE:
. . . [Our meeting last week] was the largest we have had, about 400, I should think. . . . We had one male auditor, who refused to go out when H. G. L[udlow] told him it was exclusively for ladys, & so there he sat & somehow I did not feel his presence at all embarrassing & went on just

Weld-Grimké Papers.

as 'tho he was not there. Some one said he took notes, & I think he was a Southern spy & shall not be at all surprized if he publishes us in some Southern paper, for we have heard nothing of him here....

Some friends think *I* make too many gestures, one thinking females ought to be *motionless* when speaking in public, another fearing that *other* denominations might be offended by them, because they were unaccustomed to hear women speak in public. But I think the more a speaker can yield himself entirely to the native impulses of feeling, the better, & this is just what I do....

Last 5th day I think not more than 200 were out. Sister spoke one hour on the effects on the soul, & I finished off with some remarks on the popular objection Slavery is a political subject, therefore *women* should not intermeddle. I admitted it was, but endeavored to show that women were citizens & had duties to perform to their country as well as men.... I tried to enlighten our sisters a little in their rights & duties....

PRAY FOR US—
A.E.GÉ

11

SARAH AND ANGELINA GRIMKÉ

Letter to Sarah Douglass

Newark, N.J., February 22, 1837

In this joint letter to Sarah Douglass, Sarah Grimké's earnestly religious style contrasted with Angelina's more pragmatic account of their work. However, both letters reflected the momentum that had begun to propel them forward. And both sisters expressed hope about the potential for change in the near future. Sarah Douglass headed the only academy for black girls in Philadelphia. She and her mother, Grace, were founding members of the Philadelphia Female Anti-Slavery Society. They sat on the "colored bench" in the Grimkés' Quaker meeting house.

MY DEAR SARAH:

... It is a great comfort to us that our dear sisters in Philadelphia pray for us. We need your prayers & I believe they help us. We have had the

Weld-Grimké Letters, 1:362.

privilege of attending several female prayer meetings among the colored people in New York & uniting with them in supplicating the Lord of Hosts to open a door of deliverance for our colored brethren. I have faith to believe God will answer our petitions, but sometimes my heart trembles for my country & I fear she will persist in sin until she brings down upon her guilty head the thunders of his wrath. Oh then let us pray that she may repent quickly.

Our meetings in New York have been better attended than we expected, but it is a hard place to labor in; ten thousand cords of interest are linked with the southern slaveholder. Still there are some warm-hearted abolitionists there, & I believe the fire of Emancipation will increase until our Jubilee is proclaimed.

We came to Bloomfield [N.J.] . . . last week, held two meetings with the ladies there, & they formed a society. We have had two interesting meetings also at this place [Newark] & a society was formed here; many ladies here are much engaged on behalf of the slave & this is a very important place. Southern interest is powerful; shoes & carriages, etc. made in Newark are bartered for the gold of the South, which is gotten by the unrequited Toil of the slave. Many children attended our meetings, which rejoiced us because they will soon come on the stage of action & if they are only thoro'ly abolitionized, the bastille of slavery will fall. . . .[1]

I feel deeply for thee in thy sufferings on account of the cruel & unchristian prejudice which thou hast suffered so much from. Perhaps for the present generation we can do little on this subject especially in large cities; but if we are willing to suffer, our children will reap the reward of our afflictions & toil, & we shall meet a glorious reward hereafter.

It is so much the fashion to publish anti-slavery movements that I will just mention that our friends in N.Y. think the less said at present about what we are doing the better. Let us move quietly on for a while & the two "fanatical women," as the Richmond papers call us, may, thro' divine help, do a little good. . . .

AFFY THY FRIEND
SARAH M. GRIMKÉ

MY DEAR SARAH:

. . . We feel as yet unprepared to go fully into our delightful work, because the subject of Slavery is one of such length and breadth, height & depth that our time has been, & ought to be for some time to come,

[1]Prisoners in the Bastille, the notorious French prison that symbolized despotism, were liberated by Revolutionary forces on July 14, 1789.

spent in reading on & studying it in its various bearings, I long to be fully harnessed for it, & often feel as if I had no petition to ask for at a throne of Grace, but to be made a blessing to the free & bond colored people of our land. The more I mingle with your people, the more I feel for their oppressions & desire to sympathize in their sorrows. . . .

ANGELINA E. GRIMKÉ

12

ANGELINA AND SARAH GRIMKÉ

Letter to Sarah Douglass

New York City, April 3, 1837

While organizing the upcoming convention of antislavery women, the sisters urged their African American friends to attend the meeting, noting the harsh truth that racial prejudice would make the meeting difficult for them, but encouraging them to believe that the benefits of their presence would outweigh the pain. Angelina hoped African American delegates would help draft documents related to their race. Sarah Grimké asked Sarah Douglass to help her understand the effects of racial prejudice. Sarah and Grace Douglass did attend the 1837 Anti-Slavery Convention of American Women in New York City, Grace serving as a vice president and Sarah serving on the Committee of Arrangements.

DEAR SARAH:

. . . Whenever allusion is made to that distinction which American prejudice has made between those who wear a darker skin than we do, I feel ashamed for my Country, ashamed for the church, but the time is coming when such "respect of persons" will no more be known in our land, & the children of the Lord will think no more of a difference in the color of the skin than of that of the hair or the eyes. I was very glad to hear from Sydney Ann Lewis that thy Mother & thyself tho't of coming to our Female Convention. I am very, very glad of it, and would say all I could to urge you to do so without fail. You, my dear Sisters, have a work to do in

Weld-Grimké Papers.

rooting out this wicked feeling, as well as we. You *must be willing* to come amongst us, tho' it *may be* your feelings *may* be wounded by the "putting forth of the finger," the avoidance of a seat by you, or the glancing of the eye. To suffer these things is the sacrifice which is called for at *your hands,* & I earnestly desire that you may be willing to bear these mortifications with christian meekness, gentleness & love. They will tend to your growth in grace, & will help your paler sisters *more* than anything else to overcome their own sinful feelings. Come, then, I would say, for we need your help. I cannot help hoping that a place will be found for the Fortens too. . . . If an Address to the colored people is passed by our Convention, it will be absolutely necessary that some of them should be on the Committee to examine it before it is printed.

. . . [W]e spent yesterday week in Poughkeepsie, & brother [Gerrit] Smith & ourselves had a meeting with the colored people in the evening. About 300 attended, & it was a very satisfactory meeting I believe to all parties & for the first time in my life I spoke in a promiscuous assembly, but I found that the men were no more to me then, than the women. Some of the females present were very desirous we should hold a meeting with the ladies, & we would gladly have done so, had we not expected to leave town early the next morning.

I REMAIN THY SISTER IN THE LORD
A E GRIMKÉ

MY BELOVED SISTER:
I suppose thou still attends our meeting. I feel as if the seat you occupy there is a reproach to us, & I think the Lord must send you there to be a memorial to us of our pride & our prejudice. Yet we heed it not; like many of his other lessons we let it pass unimproved. . . . I feel as if I had taken my stand by the side of the colored American, willing to share with him the odium of a darker skin, & trust, if I am permitted again to take my seat in Arch St. Mtg. House, it will be beside thee & thy dear mother. Will it be too painful for thee to give me a description of thy feelings under the effect of steel hearted prejudice? Dear Sarah, does it sink thy spirits, does it destroy thy comfort? I pray that I may feel more & more deeply for you, that thro' the grace of God my soul may be in your soul's stead. . . .

[SARAH GRIMKÉ]

13

SARAH FORTEN

Letter to Angelina Grimké

Philadelphia, April 15, 1837

Sarah Forten responded to Angelina's desire to learn about the effects of racial prejudice. Forten evaluated the positive effects of the antislavery movement as well as the negative influence of racial prejudice on her life. She also provided a cogent assessment of the colonization movement. Daughter of one of Philadelphia's wealthiest black families, Forten wrote poetry and essays for the antislavery press, and was an active member of the Philadelphia Female Anti-Slavery Society. A poem she wrote for the 1837 Anti-Slavery Convention of American Women was read to the gathering.

ESTEEMED FRIEND:

I have to thank you for the interest which has led you to address a letter to me on subject which claims so large a share of your attention. In making a reply to the question proposed by you, I might truly advance the excuse of inability; but you well know how to compassionate the weakness of one who has written but little on the subject, and who has until very lately lived and acted more for herself than for the good of others. I confess that I am wholly indebted to the Abolition cause for arousing me from apathy and indifference, shedding light into a mind which has been too long wrapt in selfish darkness.

In reply to your question — of the "effect of Prejudice" on myself, I must acknowledge that it has often embittered my feelings, particularly when I recollect that we are the innocent victims of it; for you are well aware that it originates from dislike to the color of the skin, as much as from the degradation of Slavery. I am peculiarly sensitive on this point, and consequently seek to avoid as much as possible mingling with those who exist under its influence. I must also own that *it* has often engendered feelings of discontent and mortification in my breast when I saw that many were preferred before me, who by education, birth, or worldly circumstances were no better than myself. THEIR sole claim to notice

---

Weld-Grimké Letters, 1:379.

depending on the superior advantage of being *White;* but I am striving to live above such heart burnings, and will learn to "bear and forbear" believing that a spirit of forbearance under such evils is all that we as a people can well exert.

Colonization is, as you well know, the offspring of Prejudice. It has doubtless had a baneful influence on our People. I despise the aim of that Institution most heartily, and have never yet met one man or woman of Color who thought better of it than I do. I believe, with all just and good persons, that it originated more immediately from prejudice than from philanthropy. The longing desire of a separation induces this belief, and the spirit of "this is not your Country" is made manifest by many obstacles it throws in the way of their advancement mentally and morally. No doubt but there has always existed the same amount of prejudice in the minds of Americans towards the descendants of Africa; it wanted only the spirit of colonization to call it into action. It can be seen in the exclusion of the colored people from their churches, or placing them in obscure corners. We see it in their being barred from a participation with others in acquiring any useful knowledge; public lectures are not usually free to the colored people; they may not avail themselves of the right to drink at the fountain of learning, or gain an insight into the arts and science of our favored land. All this and more do they feel acutely. I only marvel that they are in possession of any knowledge at all, circumscribed as they have been by an all powerful prejudice. Even our professed friends have not yet rid themselves of it — to some of them it clings like a dark mantle obscuring their many virtues and choking up the avenues to higher and nobler sentiments. I recollect the words of one of the one of the best and least prejudiced men in the Abolition ranks. "Ah," said he, "I can recall the time when in walking with a colored brother, the darker the night, the better Abolitionist was I." He does not say so now, but my friend, how much of this leaven still lingers in the hearts of our white brethern and sisters is oftentimes made manifest to us; but when we recollect what great sacrifices to public sentiment they are called upon to make, we cannot wholly blame them. Many, very many are anxious to take up the cross, but how few are strong enough to bear it. For our own family, we have to thank a kind Providence for placing us in a situation that has hitherto prevented us from falling under the weight of this evil; we feel it but in a slight degree compared with many others. We are not much dependent upon the tender mercies of our enemies, always having resources within ourselves to which we can apply. We are not disturbed in our social relations; we never travel far from home and seldom go to public places unless quite sure that admission is free to all; therefore we meet with none of these mortifications which might otherwise ensue. I would recommend to my colored friends to follow our example and they would be spared some very painful realities. . . .

Do you know whether the Ladies have fixed on the day for holding their Convention? Do you not think it would be best to hold it the day before the men's meeting, for most of us would be desirous to be present at both meetings. Could you not suggest this plan? There will probably be a large delegation from our society. My sisters propose going but not as Delegates. I presume there will be a sale of fancy articles there, as we were requested to send some of our work. We are all quite busy preparing something pretty and useful. Several of our schools will have specimens of work and penmanship to be sent. . . .

My Parents and Sisters unite with me in affection to you and your excellent sister.

YOURS AFFECTIONATELY
SARAH L. FORTEN

14

ANGELINA GRIMKÉ

An Appeal to the Women of the Nominally Free States

1837

Published by the national convention of antislavery women in March 1837, Angelina Grimké's second publication forcefully asserted women's right to act on behalf of emancipation, and exhorted women to incorporate their antislavery beliefs into their daily actions. Showing the confidence that flowed from her successful public speaking, she articulated a cogent written defense of women's rights.

BELOVED SISTERS:

. . . The women of the North have high and holy duties to perform in the work of emancipation — duties to themselves, to the suffering slave, to the slaveholder, to the church, to their country, and to the world at large,

---

Angelina E. Grimké, *An Appeal to the Women of the Nominally Free States, Issued by an Anti-Slavery Convention of American Women* (New York: Dorr, 1837).

and, above all to their God. Duties, which if not performed now, may never be performed at all. . . .

Every citizen should feel an intense interest in the political concerns of the country, because the honor, happiness, and well being of every class, are bound up in its politics, government and laws. Are we aliens because we are women? Are we bereft of citizenship because we are the *mothers, wives,* and *daughters* of a mighty people? Have *women* no country—no interest stakes in public weal—no liabilities in common peril—no partnership in a nation's guilt and shame?—Has *woman* no home nor household altars, nor endearing ties of kindred, nor sway with man, nor power at a mercy seat, nor voice to cheer, nor hand to raise the drooping, and to bind the broken? . . .

What then is Slavery? It is that crime, which casts man down from that exaltation where God has placed him, "a little lower than the angels," and sinks him to a level with the beasts of the field. This intelligent and immortal being is confounded with the brutes that perish; he whose spirit was formed to rise in aspirations of gratitude and praise whilst here, and to spend an eternity with God in heaven, is herded with the beasts, whose spirits go downward with their bodies of clay, to the dust of which they were made. Slavery is that crime by which man is robbed of his inalienable right to liberty, and the pursuit of happiness, the diadem of glory, and honor, with which he was crowned, and that sceptre of dominion which was placed in his hand when he was ushered upon the theatre of creation. . . .

It is gravely urged that as it is a *political subject, women* have no concernment with it; this doctrine of the North is a sycophantic response to the declaration of a Southern representative, that women have no right to send up petitions to Congress. We know, dear sisters, that the open and the secret enemies of freedom in our country have dreaded our influence, and therefore have reprobated our interference, and in order to blind us to our responsibilities, have thrown dust into our eyes, well knowing that if the organ of vision is only clear, the whole body, the moving and acting faculties will become full of light, and will soon be thrown into powerful action. Some, who pretend to be very jealous for the honor of our sex, and are very anxious that *we* should scrupulously maintain the dignity and delicacy of female propriety, continually urge this objection to female effort. We grant that it is a political, as well as a moral subject: does this exonerate women from their duties as subjects of the government, as members of the great human family? Have women never wisely and laudably exercised political responsibilities? . . .

And, dear sisters, in a country where women are degraded and brutalized, and where their exposed persons bleed under the lash—where they are sold in the shambles of "negro brokers"—robbed of their hard earnings—torn from their husbands, and forcibly plundered of their virtue and their offspring; surely, in *such* a country, it is very natural that *women* should wish to know "the reason *why*"—especially when these outrages of blood and nameless horror are practised in violation of the principles of our national Bill of Rights and the Preamble of our Constitution. We do not, then, and cannot concede the position, that because this is a *political subject* women ought to fold their hands in idleness, and close their eyes and ears to the "horrible things" that are practised in our land. The denial of our duty to act, is a bold denial of our right to act; and if we have no right to act, then may *we* well be termed "the white slaves of the North"—for, like our brethren in bonds, we must seal our lips in silence and despair. . . . *All moral beings have essentially the same rights and the same duties,* whether they be male or female. . . .

Out of the millions of slaves who have been stolen from Africa, a very great number must have been women, who were torn from the arms of their fathers and husbands, brothers, and children, and subjected to all the horrors of the middle passage and the still greater sufferings of slavery in a foreign land.[1] . . . The great mass of female slaves in the southern states are the descendants of these hapless strangers: 1,000,000 of them now wear the iron yoke of slavery in this land of boasted liberty and law. They are our countrywomen—*they are our sisters,* and to us, as women, they have a right to look for sympathy with their sorrows, and effort and prayer for their rescue. Upon those of us especially, who have named the name of Christ, they have peculiar claims, and claims which *we must answer or we shall incur a heavy load of guilt.*

Women, too, are constituted by nature the peculiar guardians of children, and children are the victims of this horrible system. Helpless infancy is robbed of the tender care of the mother, and the protection of the father. . . .

And now, dear sisters, let us not forget that *Northern* women are participators in the crime of Slavery—too many of *us* have surrendered our hearts and hands to the wealthy planters of the South, and gone down with them to live on the unrequited toil of the Slave. Too many of *us* have ourselves become slaveholders, our hearts have been hardened under the searing influence of the system, and we too, have learned to be tyrants in the school of despots. . . .

[1] "Middle passage" refers to the long voyage from Africa to the slave colonies of North and South America, during which many African captives died.

But let it be so no longer. Let us henceforward resolve, that the women of the free states never again will barter their principles for the blood bought luxuries of the South — never again will regard with complacency, much less with the tender sentiments of love, any man "who buildeth his house by unrighteousness and his chambers by wrong, that useth his neighbor's service *without* wages, and giveth him *not* for his work." . . .

Multitudes of Northern women are daily making use of the products of slave labor. They are clothing themselves and their families in the cotton, and eating the rice and the sugar, which they well know has cost the slave his unrequited toil, his blood and his tears; and if the maxim in law be founded in justice and truth, that "the receiver is *as bad* as the thief," how much *greater* the condemnation of those, who, not merely receive the stolen products of the slave's labor, but *voluntarily* purchase them, and *continually appropriate them to their own use*. . . .

In consequence of the odium which the degradation of slavery has attached to *color* even in the free states, our *colored sisters* are dreadfully oppressed here. Our seminaries of learning are closed to them, they are almost entirely banished from our lecture rooms, and even in the house of God they are separated from their white brethren and sisters as though we were afraid to come in contact with a colored skin. . . . Yes, our sisters, little as we may be willing to admit it, yet it is assuredly true, that whenever we treat a colored brother and sister in a way different from that in which we would treat them, were they white, we do virtually *reproach our Maker* for having dyed their skins of a sable hue. . . .

Much may be done, too, by sympathizing with our oppressed colored sisters, who are suffering in our very midst. Extend to them the right hand of fellowship on the broad principles of humanity and Christianity — treat them as *equals* — visit them as *equals* — invite them to cooperate with you in Anti-Slavery and Temperance, and Moral reform Societies — in Maternal Associations, and Prayer Meetings, and Reading Companies. . . . Opportunities frequently occur in travelling, and in other public situations, when your countenance, your influence, and your hand, might shield a sister from contempt and insult, and procure for her comfortable accommodations. . . . Multitudes of instances will continually occur in which you will have the opportunity of identifying yourselves with this injured class of our fellow-beings; embrace these opportunities at all times and in all places. . . . In this way, and in this way alone, will you be enabled to subdue that deep-rooted prejudice which is doing the work of oppression in the Free States to a most dreadful extent.

15

ANTI-SLAVERY CONVENTION OF AMERICAN WOMEN

Proceedings

New York City, May 9–12, 1837

Resolutions passed at this unprecedented convention reveal the public priorities of Garrisonian women, and the place that women's rights was beginning to occupy within their ranks.

Wednesday, May 10.

The Convention was called to order at 3 o'clock, P.M.

A portion of the Scriptures was read, and prayer offered.

The Committee of Arrangements made a report recommending the following subjects for the consideration of the Convention.

1. Appeal to the Women of the *nominally* Free States;
2. Address to Free Colored Americans;
3. Letter to the Women of Great Britain;
4. Circular to the Female Anti-Slavery Societies in the United States;
5. Letter to Juvenile Anti-Slavery Societies;
6. Letter to John Quincy Adams.

. . . On motion of A. E. Grimké the following resolutions were adopted: Resolved, That we regard the combination of interest which exists between the North and the South, in their political, commercial, and domestic relations, as the true, but hidden cause of the unprincipled and violent efforts which have been made, (at the North, but made in vain,) to smother free discussion, impugn the motive, and traduce the characters of abolitionists.

Resolved, That the right of petition is natural and inalienable, derived immediately from God and guaranteed by the Constitution of the United States, and that we regard every effort in Congress to abridge this sacred

Proceedings of the Anti-Slavery Convention of American Women, Held in the City of New York, May 9–12, 1837 (New York: W. S. Dorr, 1837; reprinted as *Turning the World Upside Down: The Anti-Slavery Convention of American Women, Held in New York City, May 9–12, 1837* [New York: Feminist Press, 1987]).

right, whether it be exercised by man or woman, the bond or the free, as a high-handed usurpation of power, and an attempt to strike a death-blow at the freedom of the people. And therefore that it is the duty of every woman in the United States, whether northerner or southerner, annually to petition Congress with the faith of an Esther, and the untiring perseverance of the importunate widow, for the immediate abolition of slavery in the District of Columbia and the Territory of Florida, and the extermination of the inter-state slave trade.

On motion of S. M. Grimké the following resolution was adopted:

Resolved, That we regard those northern men and women, who marry southern slaveholders, either at the South or the North, as identifying themselves with a system which desecrates the marriage relation among a large portion of the white inhabitants of the southern states, and utterly destroys it among the victims of their oppression.

The movers of the previous resolutions, sustained them by some remarks. . . .

S. M. Grimké offered the following resolution:

Resolved, That whereas God has commanded us to "prove all things and hold fast that which is good,"—therefore, to yield the right, or exercise of free discussion to the demands of avarice, ambition, or worldly policy, would involve us in disobedience to the laws of Jehovah, and that as moral and responsible beings, the women of America are solemnly called upon by the spirit of the age and the signs of the times, fully to discuss the subject of slavery, that they may be prepared to meet the approaching exigency, and be qualified to act as women, and as Christians, on this all-important subject.

The resolution was supported by the mover, A. E. Grimké, and Lucretia Mott.

A. E. Grimké offered the following resolution:

Resolved, That as certain rights and duties are common to all moral beings, the time has come for woman to move in that sphere which Providence has assigned her, and no longer remain satisfied in the circumscribed limits with which corrupt custom and a perverted application of Scripture has encircled her; therefore that it is the duty of woman, and the province of woman, to plead her cause of the oppressed in our land and to do all that she can by her voice, and her pen, and her purse, and the influence of her example, to overthrow the horrible system of American slavery.

The resolution was sustained by the mover, and by Lucretia Mott. Amendments were offered by Mary Grew and Mrs. A. L. Cox, which called forth an animated and interesting debate respecting the rights and duties of women. The resolution was finally adopted, without amendments, though ***not unanimously.***

Adjourned to Thursday morning, 10 o'clock.

Among those who voted against the adoption of this resolution, the following wished to have their names recorded in the minutes, as disapproving of some parts of it:—Mrs. Brower, Mrs. A. L. Cox, Mrs. Sophronia Johnson, Mrs. R. W. Lambden, Mrs. A. J. Lane, Mrs. R. G. Williams, Mrs. G. F. Martyn, Mrs. O. Willcox, Miss A. Rankin, Miss A. J. Dunbar, Miss H. Willcox, and Ruby Knight.

Selected Resolutions on Thursday, May 11:

On motion of Mrs. A. L. Cox, seconded by Rebecca B. Spring,

Resolved, That there is no class of women to whom the anti-slavery cause makes so direct and powerful an appeal as to ***mothers;*** and that they are solemnly urged by all the blessings of their own and their children's freedom, and by all the contrasted bitterness of the slave-mother's condition, to lift up their hearts to God on behalf of the captive, as often as they pour them out over their own children in a joy with which "no stranger may intermeddle"; and that they are equally bound to guard with jealous care the minds of their children from the ruining influences of the spirit of pro-slavery and prejudice, let those influences come in what name, or through what connections they may.

A. W. Weston offered the following resolution, viz:

Resolved, That we feel bound solemnly to protest against the principles of the American Colonization Society, as anti-Republican and anti-Christian, that we believe them to have had a most sorrowful influence in removing the chains of the slave by recognizing him as the property of his master, and in strengthening the unreasonable and unholy prejudice against our oppressed brethren and sisters, by declaring them "almost too debased to be reached by the heavenly light," that to the slave, the Society offers exile or bondage; to the free man, persecution or banishment, and that we view it as an expatriation Society.

On motion of A. E. Grimké,

Resolved, That this Convention do firmly believe that the existence of an unnatural prejudice against our colored population, is one of the chief pillars of American slavery—therefore, that the more we mingle with our oppressed brethren and sisters, the more deeply are we convinced of the sinfulness of that anti-Christian prejudice which is crushing them to the earth in our nominally Free States—sealing up the fountains of knowledge from their panting spirits, and driving them into infidelity, and that we deem it a solemn duty for every woman to pray to be delivered from such an unholy feeling, and to act out the principles of Christian equality by associating with them as though the color of the skin was of no more consequence than that of the hair, or the eyes.

16

CATHARINE E. BEECHER

Essay on Slavery and Abolitionism, with Reference to the Duty of American Females

1837

Responding to Angelina Grimké's Appeal to the Christian Women of the South *(1836), Catharine Beecher vigorously defended different forms of female power. Her goal of feminizing the teaching profession led her to promote women's power within the family. Opposing slavery, but fearful of civil war, Beecher promoted more acceptable avenues for change through slow reforms and education.*

MY DEAR FRIEND:

Your public address to Christian females at the South has reached me, and I have been urged to aid in circulating it at the North. I have also been informed, that you contemplate a tour, during the ensuing year, for the purpose of exerting your influence to form Abolition Societies among ladies of the non-slave-holding States.

Our acquaintance and friendship give me a claim to your private ear; but there are reasons why it seems more desirable to address you, who now stand before the public as an advocate of Abolition measures, in a more public manner.

The object I have in view, is to present some reasons why it seems unwise and inexpedient for ladies of the non-slave-holding States to unite themselves in Abolition Societies; and thus, at the same time, to exhibit the inexpediency of the course you propose to adopt. . . .

Catharine E. Beecher, *Essay on Slavery and Abolitionism, with Reference to the Duty of American Females* (Philadelphia: Perkins, 1837).

Now Abolitionists are before the community, and declare that all slavery is sin, which ought to be immediately forsaken; and that it is their object and intention to promote the *immediate emancipation* of all the slaves in this nation. . . . [R]eproaches, rebukes, and sneers, were employed to convince the whites that their prejudices were sinful. . . .

[T]he severing of the Union by the present mode of agitating the question . . . may be one of the results, and, if so, what are the probabilities for a Southern republic that has torn itself off for the purpose of excluding foreign interference, and for the purpose of perpetuating slavery? . . .

Heaven has appointed to one sex the superior, and to the other the subordinate station, and this without any reference to the character or conduct of either. It is therefore as much for the dignity as it is for the interest of females, in all respects to conform to the duties of this relation. . . . But while woman holds a subordinate relation in society to the other sex, it is not because it was designed that her duties or her influence should be any the less important, or all-pervading. But it was designed that the mode of gaining influence and of exercising power should be altogether different and peculiar. . . .

Woman is to win every thing by peace and love; by making herself so much respected, esteemed and loved, that to yield to her opinions and to gratify her wishes, will be the free-will offering of the heart. But this is to be all accomplished in the domestic and social circle. . . . But the moment woman begins to feel the promptings of ambition, or the thirst for power, her aegis of defence is gone. All the sacred protection of religion, all the generous promptings of chivalry, all the poetry of romantic gallantry, depend upon woman's retaining her place as dependent and defenceless, and making no claims, and maintaining no right but what are the gifts of honour, rectitude and love.

A woman may seek the aid of co-operation and combination among her own sex, to assist her in her appropriate offices of piety, charity, maternal and domestic duty; but whatever, in any measure, throws a woman into the attitude of a combatant, either for herself or others—whatever binds her in a party conflict—whatever obliges her in any way to exert coercive influences, throws her out of her appropriate sphere. . . .

If it is asked, "May not woman appropriately come forward as a suppliant for a portion of her sex who are bound in cruel bondage?" It is replied, that, the rectitude and propriety of any such measure, depend entirely on its probable results. If petitions from females will operate to exasperate; if they will be deemed obtrusive, indecorous, and unwise, by those to whom they are addressed; . . . if they will be the opening wedge, that will eventually bring females as petitioners and partisans into every political measure that may tend to injure and oppress their sex . . . then it is neither appropriate nor wise, nor right, for a woman to petition for the relief of oppressed females. . . .

In this country, petitions to congress, in reference to the official duties of legislators, seem, IN ALL CASES, to fall entirely without the sphere of female duty. Men are the proper persons to make appeals to the rulers whom they appoint, and if their female friends, by arguments and persuasions, can induce them to petition, all the good that can be done by such measures will be secured. But if females cannot influence their nearest friends, to urge forward a public measure in this way, they surely are out of their place, in attempting to do it themselves. . . .

It is allowed by all reflecting minds, that the safety and happiness of this nation depends upon having the *children* educated, and not only intellectually, but morally and religiously. There are now nearly two millions of children and adults in this country who cannot read, and who have no schools of any kind. To give only a small supply of teachers to these destitute children, who are generally where the population is sparse, will demand *thirty thousand teachers* at the moment and an addition of *two thousand every year.* Where is this army of teachers to be found? Is it at all probable that the other sex will afford even a moderate portion of this supply? . . . Men will be educators in the college, in the high school, in some of the most honourable and lucrative common schools, but the *children,* the *little children* of this nation must, to a wide extent, be taught by females, or remain untaught. . . . And as the value of education rises in the public mind . . . women will more and more be furnished with those intellectual advantages which they need to fit them for such duties.

The result will be, that America will be distinguished above all other nations, for well-educated females and for the influence they will exert on the general interests of society. But if females, as they approach the other sex, in intellectual elevation, begin to claim, or to exercise in any manner, the peculiar prerogatives of that sex, education will prove a doubtful and dangerous blessing. But this will never be the result. For the more intelligent a woman becomes, the more she can appreciate the wisdom of that ordinance that appointed her subordinate station.

But it may be asked, is there nothing to be done to bring this national sin of slavery to an end? Must the internal slave-trade, a trade now ranked as piracy among all civilized nations, still prosper in our bounds? Must the very seat of our government stand as one of the chief slave-markets of the land; and must not Christian females open their lips, nor lift a finger, to bring such a shame and sin to an end? To this it may be replied, that Christian females may, and can say and do much to bring these evils to an end; and the present is a time and an occasion when it seems most desirable that they should know, and appreciate, and *exercise* the power which they do possess for so desirable an end. . . .

In the present aspect of affairs among us, when everything seems to be tending to disunion and distraction, it surely has become the duty of every female instantly to relinquish the attitude of a partisan, in every matter of clashing interests, and to assume the office of a mediator, and an advocate of peace. And to do this, it is not necessary that a woman should in any manner relinquish her opinion as to the evils or the benefits, the right or the wrong, of any principle of practice. But, while quietly holding her own opinions, and calmly avowing them, when conscience and integrity make the duty imperative, every female can employ her influence, not for the purpose of exciting or regulating public sentiment, but rather for the purpose of promoting a spirit of candour, forbearance, charity, and peace.

REDEFINING THE RIGHTS OF WOMEN: THE GRIMKÉ SISTERS SPEAK IN MASSACHUSETTS, SUMMER 1837

17

ANGELINA GRIMKÉ

Letter to Jane Smith

Boston, May 29, 1837

In this letter Angelina reflected on her embrace of "the rights of women" in her "public labor." Amazed by her success, she began to seek support for the untrodden path she was entering. She also expressed interest in another radical notion emerging within the abolitionist movement: "non-resistance" or the repudiation of civil government.

Weld-Grimké Papers.

My Dear Jane:
. . . [At the convention of the American Anti-Slavery Society] a peace resolution was brought up, but this occasioned some difficulty, on account of non-resistance here meaning a repudiation of civil Government, & of course we cannot expect many to be willing to do this. There was no difficulty as to war itself. Indeed my own mind is all in a mist & I desire earnestly to know what is the truth about it, whether all civil government is an usurpation of Gods authority or not. Hast thou ever tho't about it, & what is thy opinion?

It has really been delightful to mingle with our brethren & sisters in this city. On 5th day evening we had a pleasant meeting of Abolitionists at Francis Jackson's, in the rooms where the Female AntiSlavery Meeting was held. On 6th day evening, we had just another such at Friend Chapman's, Ann's father. Here I had a long talk with the brethren on the rights of women & found a very general sentiment prevailing that it was time our fetters were broken. Goodell said he was well aware that women could not perform their duties as moral beings, under the existing state of public sentiment. M Child & M Chapman support the same views. Indeed very many seem to think that a new order of things is very desirable in this respect.

And now, my dear friend, in view of these things, I feel as if it is not the cause of the slave only which we plead, but the cause of woman as a responsible moral being, & I am ready to exclaim, "Who is sufficient for these things?" These holy causes must be injured if they are not helped by us. What an untrodden path we have entered upon! Sometimes I feel almost bewildered, amazed, confounded & wonder by what strange concatenation of events I came to be where I am & what I am. And if I look forward, I am no less bewildered. I see not to what point, all these things are leading me. I wonder whether I shall make shipwreck of the faith — I cannot tell — but one thing comforts me, I do feel as tho' the Lord had sent us, & as if I was leaning on the arm of my beloved. I do not believe we are going into this warfare at our own charges (spiritually), tho' I rejoice the Lord has provided for our doing so in a pecuniary point of view.

Tomorrow, we begin our public labor at Dorchester. . . . Pray for us, dear Jane. We need it *more* than ever. We see only in a glass darkly what results are to grow out of this experiment. I tremble for fear. . . . Sister is to speak at the Moral Reform Society this afternoon. I will leave this open & say something about it.

We have just returned from the meeting, & the Lord was there to help us, for I, too, opened my mouth, tho' I had refused to engage to do so.

About 300, I guess, were present & appeared interested in the remarks made. We broached one part of the subject, which I doubt not was new to many, i. e., that this reform was to begin in *ourselves.* We were polluted by it, our moral being was seared & scathed by it. Look at our feelings in the society of *men,* why the restraint & embarrassment? If we regarded each other as *moral* & intellectual beings merely, how pure & elevated & dignified would be our feelings towards, & intercourse with them. How is the solemn & sacred subject of marriage regarded & talked about? My heart is pained, my womanhood is insulted, my moral being is outraged continually, & I told them so. After we had finished, many women came up & expressed their pleasure & satisfaction at this part particularly of our remarks. They were their own feelings, but had never heard them expressed before.

No doubt, thou wilt wish to know whether the Boston women have answered our high expectations—they have: Maria Chapman, particularly, is one of the noblest women I ever saw. She has been 3 times to see us: there is real antislavery here: a heart to work, a tongue to speak. We feel ourselves surrounded by an elastic atmosphere which yields to the stroke of the wings of effort & sends up the soaring spirit still higher & swifter in its upward flight. In New York we were allowed to sit down & do nothing. Here, invitations to labor pour in from all sides. . . .

Farewell my dear Jane. May we often meet where spirits blend in prayer is the desire of Thy Angelina.

18

MARIA CHAPMAN

"To Female Anti-Slavery Societies throughout New England"

Boston, June 7, 1837

Urging women abolitionists throughout New England to support the Grimké sisters, Maria Chapman, a powerful member of the Boston Female Anti-Slavery Society, highlighted their advocacy of women's rights as an integral part of their antislavery message.

Weld-Grimké Letters, 1:395.

CHRISTIAN FRIENDS:
The purpose of this letter is to entreat, in the name of the Boston Female Anti-Slavery Society, that you will afford every facility in your power to Sarah M. and Angelina E. Grimké, for the prosecution of their labours in the cause of emancipation.

With their names and characters, with their noble sacrifices and with their published works, you are well acquainted, and therefore there is no need that we should dwell on all the circumstances growing out of these which so peculiarly fit them to dispense the truth respecting the conflicting principles of Freedom and slavery.

One thing we cannot omit to mention, which marks them eminently qualified for the promulgation of Anti Slavery principles;—the elevated and Christian point of view from which they behold the condition of woman; her duties and her consequent rights. It is of paramount importance that both men and women should understand their true positions and mighty responsibilities to this and to coming generations. In all spiritual things their functions are identical. Both are created to be parents and educators; both for all the duties growing out of that spiritual equality here and for communion with their maker during their immortal life hereafter; Neither for helplessness or dependence; neither for arbitrary dictation; each to obey the commands of God as responsible to him alone. Such is our view of the primary duties of our race. With respect to secondary pursuits, whether mercantile, mechanical, domestic, or professional—the machinery of mortal existence—"the tools to whosoever can use them." All are alike bound to the strenuous exercise of such faculties as God has given them.

We could not confidently commend to your hearts, or receive unreservedly to our own, any who were grinding in the narrow mill of a corrupt publick opinion on this point; but in view of the justness of their theory and the faithfulness of their practice concerning it, we earnestly entreat you in the words of Paul, "help THESE women, who have laboured thus in the Gospel," and thereby help us to manifest gratitude for the important aid they are affording to Anti slavery enterprise. Help them to exalt the national character of our women—so inferior to that of the Maternal Ancestry who in 1620 "fled from their spheres in England, and journeyed here with their little ones," shelterless in the wintry air, that they might pursue their christian course unimpeded by sneers or ridicule, ecclesiastical mandates or publick outrage.

Let us help one another to refute the idea, that while the chief end of *man* is to glorify God and enjoy him forever, woman is sharer of the like glorious destiny "but as it were in sort [in part], or limitation."

Dear friends, let us urge on you the importance of making available to the cause of Freedom the scattered energies of your respective neighborhoods by gathering together and seeking the cooperation of all whose interest in suffering humanity is leading them to ask "what shall we *do*?" The numbers of such, in every place, are small in comparison with those who will undertake to dictate to *you* what you shall *not* do.

We are not entirely without experience. Trust us when we say that we have found those the most effectual helpers who come to us least encumbered by the trappings of this world, and unfettered from the thraldom of its ways.

Let there be no exclusive system adopted in our societies. Ask no one's sect, rank or colour. Whosoever *will,* let them come. If our worship be sincere of the God who created our race free, and the Savior who came to redeem them from bondage, it will so appear in our active exertions for our enslaved countrymen, that the selfish, the hypocritical and the unfaithful, will be compelled to hold themselves aloof from our ranks. There is no danger to be apprehended from the companionship of any others, for a holy cause purifies the heart, and refines and exalts the ideal of all who embrace it in sincerity.

We renewedly commend to you these our beloved friends, nothing doubting that they will receive from you that hospitality of the heart which will be to them an assurance that they have not consecrated their lives and fortunes to the cause of Christian Freedom in vain.

IN BEHALF OF THE BOSTON FEMALE ANTI-SLAVERY SOCIETY.
MARIA WESTON CHAPMAN. SEC.

19

ANGELINA GRIMKÉ

Letter to Jane Smith

Danvers, Mass., June 1837

By June the sisters were speaking almost every day to large audiences, most of which included men as well as women. They had become highly skilled professionals.

I do not know, My Beloved Jane, why it is that I have not heard from you since I left New York, but I am sure that thou wantest to hear something of our getting along since my last, written from Boston near three weeks ago. I will, therefore, copy a leaf from our day book since the time I wrote. 7th of 6th month. Spoke before the Anti Slavery Society in Boston, in Washington Hall, where they had been mobbed 18 months before.[1] About 400 present; many could not get in. 5 life members. 33 annual subscribers. 8th. Held a meeting at Brookline, in the house of Saml Philbrick. 75 present, first A S meeting ever held in the town. much opposition to be felt & very hard to speak to such strong hearts. 9th. Addressed the A S Society at North Weymouth, about 120 present—great apathy—hard to speak. 9 new subscribers—near 30 men present. . . . 16th. Attended a Peace meeting in the Vestry of the Old South Ch[urch] in Boston. about 250 out. took the *ultra* ground, on law & civil government also. 18th. Addressed the A S Sy of So Weymouth. about 150 out. tho' it was very rainy. 19th. A S Sy in Boston. 550 women. 50 men. very easy to speak because there was great openness to hear. about 50 new subscribers added. 21st. Attended the Anniversary of the Lynn A S Sy. Spoke on the Report & Resolutions—about 500 women present.

In the evening of the same day, addressed our first large mixed audience. about 1000 present. Great openness to hear & ease in speaking. 22d. Held another in Lynn, but in a smaller house, so that it was crowded to excess. about 600 seated. many went away, about 100 stood around

[1] Refers to a mob that in October 1835, dragged William Lloyd Garrison out of a meeting of the Boston Female Anti-Slavery Society and threatened to hang him.

Weld-Grimké Papers.

the door, & we were told that on each window on the outside stood three men with their heads above the lowered sash. very easy speaking indeed. 23d. Held a meeting here [Danvers]. about 200 out — a few of the brethren. very hard speaking: I gave them a complete scolding, which I afterwards found they deserved. Think it likely we shall not have much more than half as many this afternoon in consequence of it, but the truth must be told.

And now thou will want to know how we feel about all these things.... Whilst in the act of speaking I am favored to forget little "I" entirely & to feel altogether hid behind the great cause I am pleading. Were it not for this feeling, I know not how I could face such audiences without embarrassment.

It is wonderful to us how the way has been opened for us to address mixed audiences, for most sects here are greatly opposed to public speaking for women, but curiosity in many & real interest in the AS cause in others induce the attendance of our meetings. When they are over, we feel as if we had nothing to do with the results. We cast our burden upon the Lord, & feel an inexpressible relief until the approach of another meeting produces an exercise & sense of responsibility which becomes at times almost insupportable. At some of the meetings I have really felt sick until I rose to speak. But our health has been good & we bear the exertion of body & exercise of mind wonderfully. Our compass of voice has astonished us, for we can fill a house containing 1000 persons with ease.... It seems that the Salem meeting house was granted for us... so that we are to speak in it on 2nd day afternoon. I almost feel sorry for I am afraid I shall feel the influence of Quaker restrictions & be ill at ease.... Our headquarters [in Newburyport] is to be with Henry C. Wright, one of the best men I ever met with & there he says we must rest, but if we continue to bear one meeting a day as well as we have done, we shall not want to rest at all.

Hast thou read CE Beecher's book? I am answering it by letter in the Liberator & requested that they might be sent to thee by post. I have not spared her at all, as thou wilt perceive. It was one of the most subtle things I ever saw.... I do not know how I shall find language strong enough to express my indignation at the view she takes of the woman's character & duty....

Sister enjoys more real comfort of mind than I ever saw her enjoy before & it is delightful to be there yoked with her in this work, but we often wish we could divide ourselves in order to do far more work than we can at present....

THY EVER AFFECTIONATE A E GRIMKÉ

20

ANGELINA GRIMKÉ

Letter to Jane Smith

New Rowley, Mass., July 25, 1837

Angelina's increasing radicalism on the question of civil government accompanied her growing militancy on women's rights.

MY DEAR JANE:

I am truly glad thou wilt have an opportunity of becoming acquainted with brother Wright. He will tell thee all about his views of civil government & be not afraid to be converted. I can truly say that until I embraced them I never understood the full extent of that Liberty wherewith Christ makes his followers *free*. It is indeed delightful to realize that He is our King, our lawgiver & our judge. Without these views I know not *how* I could press forward in the path of difficulty which lies before me. . . .

Some of these places are only villages, so that the few hundred who have come out have been a good many for the size of them. But our *womanhood*— it is as great offense to some as our Abolitionism. I will let H C W [Henry Clarke Wright] tell thee what a war is waged against it. The whole land seems roused to discussion on the *province of woman*, & I am glad of it. We are willing to bear the brunt of the storm, if we can only be the means of making a breach in the wall of public opinion, which lies right in the way of woman's true dignity, honor & usefulness. Sister Sarah does preach up woman's rights most nobly & fearlessly, & we find that many of our New England sisters are ready to receive these strange doctrines, feeling as they do, that our whole sex needs an emancipation from the thraldom of public opinion. What doest thou think of some of them walking 2, 4, 6 & 8 miles to attend our meetings?

But I must forbear — as Sisters voice failed her on account of cold. I have had to bear the brunt of the meeting for two days, speaking an hour & a half today & an hour and three quarters yesterday & as it is ½ past 9 I must say Fare thee well to night my dear friend.

Weld-Grimké Papers.

21

SARAH AND ANGELINA GRIMKÉ

Letter to Amos Phelps

Groton, Mass., August 3, 1837

Clergymen within the Garrisonian ranks criticized the sisters for diverting their energy to women's rights. To them the sisters vigorously defended their tactical decision to defend women's rights now rather than later. Amos Phelps, a Congregational minister and an agent of the American Anti-Slavery Society, had asked them to stop lecturing before mixed audiences and to allow him to publish his request and their agreement.

DEAR BROTHER PHELPS:

Thy letter, which we received yesterday at Lowell, neither surprised us nor moved us, because we are prepared to find opposition & to meet with condemnation from the ministry *generally*. . . . The clergy have done an infinite injury to woman, & woman in the coming conflict will, we apprehend, be much in the situation of Paul, when he said: "No man stood by me." But, my brother, we have planted our feet on the Rock of Ages, & our trust is in Him who saith, "Trust in the Lord for in the Lord Jehovah is everlasting strength." Our views & principles & practices in this matter are founded upon the immutable Truth of God, & we believe that to abandon them would to be to surrender our rights as moral & responsible beings. . . .

Thou sayest our present course makes the Anti-Slavery cause responsible for what, in thy judgement, we should not make it responsible. We do not, & we cannot surrender our moral accountability to any society, & when we united ourselves to join the A. S. S., we did not give up our free agency. I can, therefore, only repeat what I said on the first hand, that if in the performance of duty any reproach is cast on the A. S. cause thro' our instrumentality, I do not think it is our fault, however much we may regret it. To close the doors *now* against our brethren wd. be a violation of our fundamental principle that man & woman are created equal, & have the same duties & the same responsibilities as moral beings. If,

---

Boston Public Library, Ms. A.21.7 (31).

therefore, it is right for thee, my dear brother, to lecture to promiscuous assemblies, it is right for us to do the same. . . .

We hold no commission from the Quakers to do what we are doing, nor do we in the least defend our present course by sheltering ourselves under our Quakerism. If it is wrong for us to speak the Truths of the Gospel in mixed assemblies, our belonging to the So. of Fds. does not make it right. We, therefore, always disclaim this reason & express ourselves as acting from a conviction of duty based on the Scriptures. This subject is worthy of candid & *prayerful* investigation, & we hope that for thy own sake thou wilt be willing to examine it. We should regret that a brother whom we esteem so highly should identify himself with the men who sent forth the Pastoral Letter, a letter which aims to tear from the woman her dearest rights & substitute the paltry privilege of leaning upon the fallen creature instead of the strong arm of the Almighty God. We believe that this subject of women's rights & duties must come before the public for discussion, so that the Lord will help us to endure the opposition, contumely & scorn which will be cast upon womanhood & that he will make us more than conquerors thro' him that loved us. We would, therefore, entreat our brethren to stand still not for our sakes, but for their own, lest peradventure that be found fighting against God. . . .

As far as we are concerned, we are entirely willing that thou shouldst "*say publicly*" any thing thou wishes . . .

THY SISTER IN THE BONDS OF THE GOSPEL
SARAH M. GRIMKÉ

22

Pastoral Letter: The General Association of Massachusetts to Churches under Their Care

July 1837

Issued by the highest authority in the Congregational Church, the most powerful church in Massachusetts, this letter rebuked the Grimkés' assertion of women's rights with words like "permanent injury," "shame and dishonor," and "degeneracy and ruin."

New England Spectator, July 12, 1837.

Brethren and Friends,—Having assembled to consult upon the interests of religion within the Commonwealth, we would now, as pastors and teachers, in accordance with the custom of this Association, address you on some of the subjects which at the present time appear to us to have an important bearing upon the cause of Christ. . . .

We invite your attention to the dangers which at present seem to threaten the female character, with wide spread and permanent injury.

The appropriate duties and influence of women are clearly stated in the New Testament. Those duties and that influence are unobtrusive and private, but the sources of mighty power. When the mild, dependent, softening influence of woman upon the sternness of man's opinion is fully exercised, society feels the effects of it in a thousand forms. The power of woman is in her dependence, flowing from the consciousness of that weakness which God has given her for her protection, and which keeps her in those departments of life that form the character of individuals and of the nation. There are social influences which females use in promoting piety and the great objects of Christian benevolence which we cannot too highly commend. We appreciate the unostentatious prayers and efforts of woman in advancing the cause of religion at home and abroad; in Sabbath schools; in leading religious inquirers to the pastors for instruction; and in all such associated efforts as becomes the modesty of her sex; and earnestly hope that she may abound more and more in these labors of piety and love.

But when she assumes the place and tone of man as a public reformer, our care and protection of her seem unnecessary; we put ourselves in self-defence against her; she yields the power which God has given her for protection, and her character becomes unnatural. If the vine, whose strength and beauty is to lean upon the trellis work and half conceal its clusters, thinks to assume the independence and the overshading nature of the elm, it will not only cease to bear fruit, but fall in shame and dishonor into the dust. We cannot, therefore, but regret the mistaken conduct of those who encourage females to bear an obtrusive and ostentatious part in measures of reform, and countenance any of that sex who so far forget themselves as to itinerate in the character of public lecturers and teachers.—We especially deplore the intimate acquaintance and promiscuous conversation of females with regard to things which ought not to be named; by which that modesty and delicacy which is the charm of domestic life, and which constitutes the true influence of woman in society is consumed, and the way opened, as we apprehend, for degeneracy and ruin.[1] We say these things, not to discourage proper influence against sin, but to secure such reformation as we believe is Scriptural, and will be permanent.

[1] "Things which ought not to be named" referred to the sisters' discussions of sexual relations (both forced and consensual) between masters and female slaves. In this phrase the Pastoral Letter also indirectly denounced the practice by which Female Moral Reform Society members publicized the names of men who frequented prostitutes.

23

Lecture by Albert Folsom, Pastor, Universalist Church

Hingham, Mass., August 27, 1837

Ministers like Folsom tried to rally respectable opinion against abolitionism by tainting it with the advocacy of women's rights. By printing such attacks,* The Liberator *kept readers informed about the cultural debates raging around them.

. . . The legitimate effect of being converted to the popular measures of the Abolitionists, (popular, I mean, among a certain class — not with the great mass of the people, — God forbid) is a neglect of some of the appropriate duties of woman. . . .

If it is not permitted unto women to speak publicly upon the subject of religion, it verily is no part of their right or privilege to be heard upon the subject of slavery. If it is a shame for a woman to speak in the church upon one topic, it is no less shameful for her to raise her voice upon any other theme. And in all instances of the kind, females go counter to the established opinion of the world and the express commands of Holy Writ. Hence, they ought to be looked upon as "busy bodies" "speaking things which they ought not."

"The simplicity of Christ" peremptorily forbids those practices, to which we have alluded, as it does all interference in the concerns of the state, on the part of the female portion of the community. It is unbecoming the dignity of the feminine class of society to importune the National

"Abolition Women," *The Liberator*, Sept. 22, 1837.

Court, year after year, upon the difficult subject of slavery. Still more irreverent and unbecoming is it to threaten incessant application, until Congress shall grant the stale prayer of the misguided petitioners, who are made up of all classes, characters and colors.

From such improprieties, may reason and good sense deliver you all. May a suitable regard to your own characters and sex deter you from entering upon the inappropriate and unlawful duties of public life, or from seeking unenviable notoriety after the way and manner of some.

24

ANGELINA GRIMKÉ

Letter to Jane Smith

Groton, Mass., August 10, 1837

Pushing ahead despite growing opposition to the sisters' women's rights stance and to their lecturing before men as well as women, the Grimké's often encountered hostile audiences, especially at Andover, seat of New England's most prominent theological seminary. But at Lowell, they attracted 1,500, including a large number of working girls and women, and spoke with great success beneath the blazing chandeliers of the city hall.

MY DEAR JANE:

. . . I wonder not that some think there is danger of my thinking "more highly of myself than I ought to think." There is danger, & therefore I believe the Lord is mercifully preparing an opposition to our labor which I trust will tend to humble & keep us at the foot of the Cross. No doubt H C W [Henry Clarke Wright] told you that a storm was gathering all around against our *womanhood;* the Ministers especially are in great trepidation, & I should not be at all surprised, if in 3 months, almost every pulpit was closed against us. . . .

My only fear is that some of the anti-slavery brethern will commit themselves, in this excitement, against *woman's rights & duties,* before

Weld-Grimké Papers.

they examine the subject, & will, in a few years, regret the steps they may take. This will soon be an absorbing topic.

It must be discussed whether women are moral & responsible beings, and whether there is such a thing as *male and female virtue & male and female* duties &c. My opinion is that there *are none* & that this false idea has driven the plowshare of ruin over the whole field of morality. My idea is that whatever is morally right for a man to do is morally right for a woman to do. I recognize no rights but human rights. I know nothing of men's rights and women's rights; for in Christ Jesus there is neither male nor female. . . . I am persuaded that woman is not to be as she has been, a mere secondhand agent in the regeneration of a fallen world, but the acknowledged equal and co-worker with man in this glorious work. . . .

Hubbard Winslow[1] of Boston has preached & published a sermon to set forth the *proper sphere of our sex,* which I think I shall review when I can get time. I am truly glad that men are not ashamed to come out boldly & tell us just what is in their hearts.

But I must take up the account of our labors now as I know you want to hear about them. On the 28th we had our second meeting at Andover, a heavy thunderstorm was coming up when we went but the meeting house was full, about 800 present, a good many of the students were out. I never felt as if I was speaking before such a formidable array of talent & learning & prejudice against my womanhood. I felt miserable in view of the meeting, but the Lord helped me through. I spoke on these subjects, the South never was preparing for Emancipation, therefore we could not have rolled it back & the effects of Abolitionism on the South. 29th went to Methuen in the rain, felt pretty nearly worn out, things looked gloomy, did not feel much like speaking that evening. After dinner the weather cleared off beautifully—about 1000 people were out, the largest audience it was that had ever been collected there. Remarks desultory. Rested on the Sabbath, staying at home as usual. 31st went to Lowell in the afternoon thinking it was only a manufacturing place. I thot the meetings would not be of very great consequence, but found to my surprise that it contained 15,000 inhabitants & that the friends expected a large meeting & had engaged the *city hall.* I was fairly frightened when I found myself in a city audience of 1500, surrounded with a blaze of light from chandeliers & lamps. I hardly knew what would

[1]Hubbard Winslow (1799–1864) was a Congregational minister who denounced women abolitionists in a sermon that he later expanded into a book, *The Appropriate Sphere of Woman* (Boston: Jordan Weeks & Co., 1837).

become of me. Sister says I looked just before I rose, as tho' I was saying to myself, the time has arrived & the *sacrifice must be offered,* & I felt just so. The Lord stood at my right hand & sustained & carried me through. . . .

On the 2d came to this lovely little village [Groton]. . . . Here Anne Weston of Boston had come to meet us. She says the Boston women will stand by us in the contest for woman's rights, that they were very glad to find we had accepted the challenge of a discussion at Amesbury,[2] on account of its bearings on the province of *woman &c.*

On the 3d brother Stanton came here, found he was sound on the subject of woman's rights. He went to meeting with us in the evening, opened it with a precious prayer & sat with us in the pulpit. About 500 out, the largest Anti Slavery audience that has yet attended in this place. He says he wants very much so to arrange some meeting, so that *we & he* may speak at it together. This would be an *irretrievable commitment,* but I doubt whether the time has fully come for such an anomaly in Massachusetts. . . .

MOST AFFY WITH SISTERS LOVE TO ALL OF YOU.
A E GÉ

[2] At Amesbury the sisters had agreed to debate two men who challenged their views on slavery.

25

ANGELINA GRIMKÉ

Letter to Theodore Weld

Groton, Mass., August 12, 1837

Angelina's letter to Theodore crossed his letter to her in the mail. She expressed her anxieties about what steps the AASS leadership might take against her and Sarah. Acknowledging that their advocacy of women's rights affected men in "the tenderest relations of life," she invited Weld to tell her his views on equality within marriage.

Weld-Grimké Letters, 1:414.

MY DEAR BROTHER:
No doubt thou hast heard by this time of all the fuss that is now making in this region about our stepping so far out of the bounds of female propriety as to lecture to promiscuous assemblies. My auditors literally sit some times with "mouths agape and eyes astare," so that I cannot help smiling in the midst of "rhetorical flourishes" to witness their perfect amazement at hearing a woman speak in the churches. I wish thou couldst see Brother Phelp's letter to us on this subject and sister's admirable reply. I suppose he will soon come out with a conscientious protest against us. I am waiting in some anxiety to see what the Executive Committee mean to do in these troublous times, whether to renounce us or not.

But seriously speaking, we are placed very unexpectedly in a very trying situation, in the forefront of an entirely new contest — a contest for the *rights of woman* as a moral, intelligent & responsible being. Harriet Martineau says "God & man know that the time has not come for women to make their injuries even heard of";[1] but it seems as tho' it had come *now* & that the exigency must be met with the firmness & faith of woman in by gone ages. I cannot help feeling some regret that this shld have come up *before* the AntiSlavery question was settled, so fearful am I that it may injure that blessed cause, & then again I think this must be the Lord's time & therefore the *best* time, for it seems to have been brought about by concatenation of circumstances over which we had no control. The fact is it involves the interests of every minister of our land, & therefore they will stand almost in a solid phalanx against woman's rights, & I am afraid the discussion of this question will divide in Jacob and scatter in Israel;[2] it will also touch every man's interests at home, in the tenderest relation of life; it will go down into the very depths of his soul & cause great searchings of heart. I am glad H Winslow of Boston has come out so boldly & told us just what I believe is in the hearts of thousands of men in our land.

I must confess my womanhood is insulted, my moral feelings outraged when I reflect on these things, & I am sure *I know just* how the free colored people feel towards the whites when they pay them more than common attention; it is *not paid as a RIGHT, but given as a BOUNTY.* . . . There is not one man in 500 who really understands what

[1] Harriet Martineau (1802–1876), a British abolitionist and author of *Society in America* (1837), described the struggles of American women abolitionists in "The Martyr Age in the United States of America," in *The London and Westminister Review,* Dec. 1838.

[2] Genesis 49:7.

kind of attention is alone acceptable to a woman of pure & exalted moral & intellectual worth. Hast thou read Sister's letters in the Spectator?[3] I want thee to read them and let us know what thou thinkest of them. That a wife is not to be subject to her husband in any other sense than I am to her or she to me, seems to be strange and *alarming* doctrine indeed, but how can it be otherwise unless *she surrenders her moral responsibility,* which *no woman has a right* to do? . . .

WHO will stand by woman in the great struggle? As to our being Quakers being an excuse for our speaking in public, we do *not* stand on this ground at all; we ask no favors for ourselves, but *claim* rights for our *sex.* If it is wrong for woman to lecture or preach then let the Quakers give up their false views, and let the other sects refuse to hear their women, but if it is *right* then let *all* women who have gifts "mind their calling" and enjoy "the liberty wherewith Christ hath made them free," in that declaration of Paul, "in Christ Jesus there is neither male nor female." O! if in our intercourse with each other we realized this great truth, how delightful, ennobling and dignified it would be, but as I told the Moral Reform Society of Boston in my address, *this* reformation *must begin with ourselves.* . . .

Yesterday the sabbath, rode 12 miles to lecture at Boxboro, brother Gross having written us a pressing invitation to come and plead the cause of God's perishing poor in *his pulpit.* It so happened that yesterday was the only day we could do so before we left for Boston. Found his meeting crammed to overflowing. O! what a feeling, to see such a congregation waiting for the words that shall fall from MY unworthy lips. Thou knowest it dear brother, and can understand all about it *except that I am a woman.* I spoke an hour and a half and then stopped and took some refreshment with his family. He says J. Woodbury[4] is against our womanhood and that as *all* the Congregational ministers except himself (about here I mean) are opposed, he expects to have to fight a battle at their next meeting; and that he means to throw down the gauntlet about women's preaching. We pointed out some texts he had not tho't of and tried to throw our views before his mind. May the Lord open his heart more and more on this subject and sustain him in the sore conflict he will have to wage, if he is faithful in pleading for woman's essential rights.

I have no doubt that posterity will read withal *women* were *not* permitted to preach the gospel, with as much amazement and indignation

[3] Sarah Grimke's letters were first published in the *New England Spectator,* and then published as a book, *Letters on the Condition of Woman and the Equality of the Sexes* (Boston: Knapp, 1838).

[4] James Woodbury (1803–1861) was a Congregational minister in Connecticut who had trained with the Grimkés in New York.

as we do that no *colored* man in No. Ca[rolina] is allowed this *holy right.* Now we want thee to sustain us on the high ground of MORAL RIGHT, *not* of Quaker peculiarity. This question must be met now; let us do it as *moral* beings, & not try to turn a SECTARIAN *peculiarity* to the best account for the benefit of Abolitionism. We do not stand on Quaker ground, but on Bible ground & *moral right.* What we claim for ourselves, we claim for *every* woman who God has called & qualified with gifts & graces. Can't *thou* stand *just here* side by side with us? . . .

Mary Parker sent us word that the Boston women would stand by us if *every* body else forsook us. A[nne] Weston has been here with us & is very strong. . . .

THY SISTER IN THE BONDS OF WOMAN AND THE SLAVE.
A E GÉ

26

THEODORE WELD

Letter to Sarah and Angelina Grimké

August 15, 1837

Declaring his unqualified support of women's rights, Weld urged the sisters to put the antislavery cause first. Any women could defend women's rights, he insisted. As southerners, they were needed and in a powerful position to speak for the slave.

MY DEAR SISTERS:

. . . As to the *rights* and *wrongs* of women, it is an old theme with me. It was the *first* subject I ever *discussed.* In a little debating society when a boy, I took the ground that *sex* neither *qualified* nor *disqualified* for the discharge of any functions mental, moral, or spiritual; that there is no reason why *woman* should not make laws, administer justice, sit in the chair of state, plead at the bar or in the pulpit, if she has the qualifications, just as much as tho she belonged to the other sex. Further, that the proposition of marriage may with just the same propriety be made by

Weld-Grimké Letters, 1:425.

the ***woman*** as the ***man,*** and that the existing usage on that subject, pronouncing it ***alone*** the province of the ***man,*** and ***indelicacy*** and almost, if not quite ***immoral*** for ***woman*** to make the first advances, overlooks or rather *perverts* the sacred design of the institution and debases it into the mire of earthliness and gross sensuality, smothering the spirit under the flesh. . . . [W]e *fully agree in principle.* . . .

Now notwithstanding this, I do most deeply regret that you have begun a series of articles in the Papers on the rights of woman. Why, my dear sisters, the best possible advocacy which you can make is just what you are making day by day. Thousands hear you every week who have all their lives held that woman must not speak in public. Such a practical refutation of the dogma as your speaking furnishes has already converted multitudes. . . . Besides you are *Southerners,* have been slaveholders; your dearest friends are all in the sin and shame and peril. All these things give you great access to northern mind, great *sway* over it. You can do ten times as much on the subject of *slavery* as Mrs. Child or Mrs. Chapman. Why? Not because your powers are superior to theirs, but because you are *southerners.* You can do more at convincing the North than twenty *northern* females, tho' they could speak as well as you. Now this peculiar advantage you *lose* the moment you take *another* subject. . . . *Any* women of your powers will produce as much effect as you on the north in advocating the rights of *free* women (I mean in contradistinction to *slave* women). . . . Let us all *first* wake up the nation to lift millions of slaves of both sexes from the dust, and turn them into MEN and then . . . it will be an easy matter to take millions of females from their knees and set them on their feet. . . . All our opposers . . . will chuckle if only a part of your energies . . . can be diverted into one which will make you so obnoxious as to cripple your influence on the subject of slavery. . . .

YOUR BROTHER T. D. WELD

27

JOHN GREENLEAF WHITTIER

Letter to Angelina and Sarah Grimké

New York City, August 14, 1837

Calling the women's rights issue a "paltry grievance" when compared to needs of slaves, Whittier wrote from the head office of the American Anti-Slavery Society. He supported their right to speak to mixed audiences, but expressed anxiety about the "startling opinions" that were emerging within the movement, and urged them not to divert their energies into writing about women's rights.

MY DEAR SISTERS:

. . . I am anxious, too, to hold a long conversation with you on the subject of war, human government, and church and family government. The more I reflect on this subject, the more difficulty I find, and the more decidedly am I of the opinion that we ought to hold all these matters far aloof from the cause of abolition. Our good friend, H. C. Wright, with the best intentions in the world, is doing great injury by a different course. He is making the anti-slavery party responsible in a great degree, for his, to say the least, startling opinions. . . . But let him keep them distinct from the cause of emancipation. This is his duty.

In regard to another subject, "*the rights of woman,*" you are now doing much and nobly to vindicate and assert the rights of woman. Your lectures to crowded and promiscuous audiences on a subject manifestly, in many of its aspects, *political,* interwoven with the framework of the government, are practical and powerful assertions of the right and the duty of woman to labor side by side with her brother for the welfare and redemption of the world.

Why, then, let me ask, is it necessary for you to enter the lists as controversial writers on this question? Does it not *look,* dear sisters, like abandoning in some degree the cause of the poor and miserable slave, sighing from the cotton plantations of the Mississippi, and whose cries and groans are forever sounding in our ears, for the purpose of arguing

---

Weld-Grimké Letters, 1:423.

and disputing about some trifling oppression, political or social, which we may ourselves suffer? Is it not forgetting the great and dreadful wrongs of the slave in a selfish crusade against some paltry grievance of our own? . . . The Massachusetts Congregational Association can do you no harm if you do not allow its splenetic and idle manifesto to divert your attention from the great and holy purpose of your souls. . . .

YOUR FRIEND AND BROTHER,
JON. G. WHITTIER

28

ANGELINA GRIMKÉ

Letter to Theodore Dwight Weld and John Greenleaf Whittier

Brookline, Mass., August 20, 1837

Seeing herself in the ancient role of a prophet overturning the power of priests, Angelina scathingly denounced the clerical origins of the power arrayed against her vindication of women's rights. She saw through the strategy of silencing women, and addressed the issue of men's power, their fear of losing power, and their desire to keep it. She also defended the "rain bow" of "moral reformations" emerging within the antislavery movement, and continued her colloquy with Weld on the topic of courtship and marriage.

TO THEODORE D. WELD AND J.G. WHITTIER
BRETHREN BELOVED IN THE LORD:
As your letters came to hand at the same time & both are devoted mainly to the same subject, we have concluded to answer them on one sheet & jointly. You seem greatly alarmed at the idea of our advocating the *rights of woman*. . . . These letters have not been the means of *arousing* the public attention to the subject of Woman's rights; it was the Pastoral Letter which did the mischief. The ministers seemed panic struck at once & commenced a most violent attack upon us. I do not say *absurd*, for in truth if it can be fairly established that women *can lecture*, then why may they not preach, & if *they* can preach, then woe! woe be unto the Clerical

Weld-Grimké Letters, 1:427.

Domination which now rules the world under the various names of Genl Assemblies, Congregational Associations, etc. *This Letter,* then roused the attention of the whole country to inquire what *right* we had to open our mouths for the dumb; the people were continually told "it is a *shame* for a *woman* to speak in the churches. Paul suffered not a *woman to teach* but commanded *her* to be in silence. The pulpit is too *sacred a place for woman's* foot &c."

Now, my dear brothers, *this invasion of our rights* was just such an attack upon *us,* as that made upon Abolitionists generally, when they were told a few years ago that *they had no right* to discuss the subject of Slavery. Did *you* take no notice of this assertion? Why no! With one heart & one voice, you said, *We will* settle *this right before* we go one step further. *The time* to assert a right is *the* time when *that* right is denied. *We must establish this right,* for if we do not, it will be impossible for *us* to go on *with the work of Emancipation. . . .*

You certainly *must* know that the leaven which the ministers are so assiduously working into the minds of the people *must* take effect in process of time, & *will close every church to us,* if we give the community no reasons to counteract the sophistry of priests & levites. In this State, particularly, there is an utter ignorance on the subject. Some few noble minds bursting thro' the trammels of educational prejudice FEEL that woman does stand on the same platform of human rights with man, but even these cannot sustain their ground by argument, & as soon as they open their lips to assert her *rights,* their opponents throw perverted scripture into their faces & call O yea, clamor for proof, PROOF, PROOF! & this *they cannot* give & are beaten off the field in disgrace. Now, we are confident that there are scores of such minds panting after light onto this subject: "the children *ask* bread & no MAN giveth it unto them." There is an eagerness to understand our views. Now, is it wrong to give those views in a series of letters in a paper NOT devoted to Abolition?

And can you not see that women *could* do, & *would* do a hundred times more for the slave if she were not fettered? Why! we are gravely told that we are out of our sphere even when we circulate petitions; out of our "appropriate sphere" when we speak to women only; & out of them when we *sing* in the churches. Silence is *our* province, submission *our* duty. If, then, we "give *no reason* for the hope that is in us," that we have *equal rights* with our brethren, how can we expect to be permitted *much longer to exercise those rights?* IF I know in my own heart, I am NOT actuated by any selfish considerations . . . but we are actuated by the full conviction that if we are to do any good in the Anti Slavery cause, our *right* to labor in it *must* be firmly established; *not* on the ground of

Quakerism, but on the firm basis of human rights, the Bible. Indeed, I contend brethren that *this* is not *Quaker* doctrine; it is no more like *their* doctrine on Women than our Anti Slavery is like their Abolition—just about the same difference. I will explain myself. Women are regarded as equal to men on the ground of *spiritual gifts, not* on the broad ground of *humanity.* Woman may *preach;* this is a *gift;* but woman must not make the discipline by which *she herself* is to be governed.

O that you were here that we might have a good long, long talk over matters and things; then I could explain myself far better, & I think we could convince you that we cannot push Abolitionism forward *until* we take up the stumbling block out of the road. We cannot see with brother Weld in this matter. We acknowledge the excellence of his reasons for urging us to labor in this cause of the Slave, our being Southerners, &c. But then we say how can we expect to be able to hold these meetings much longer, when people are so diligently taught to *despise* us for thus stepping out of the "sphere of woman"!

Look at this instance: after we had left Groton, the *Abolition* minister there, at a Lyceum meeting, poured out his sarcasm & ridicule upon our heads, & among other things said, he would as soon be caught robbing a hen roost as encouraging a woman to lecture. Now, brethren, if the leaders of the people thus speak of our labors, *how long* will we be allowed to prosecute them? Answer me this question. You may depend on it, tho' to meet *this* question *may appear* to be turning out of our road, that *it is not.* IT IS NOT: we must meet it & meet it *now* & meet it like *women* in the fear of the Lord. . . . If we dare to stand upright & do our duty according to the dictates of *our own* consciences, why then we are compared to Fanny Wright, &c.

Why, my dear brothers, can you not see the deep laid scheme of the clergy against us as lecturers? They know full well that if they can persuade the people it is a *shame* for us to speak in public, & that every time we open our mouths for the dumb we are breaking a divine command, that even if we spoke with the tongues of *men* or angels, we should have no *hearers.* They are springing a deep mine beneath our feet, & we shall very soon be compelled to retreat for we shall have no ground to stand on. If we surrender the right to *speak* to the public this year, we must surrender the right to petition next year & the right to write the year after &c. What *then* can *woman* do for the slave, when she herself is under the feet of man & shamed into *silence?* Now we entreat you to weigh candidly the *whole subject,* & then we are sure you will see this is no more than an abandonment of our first love than the effort made by Anti Slavery men to establish the right of *free* discussion.

With regard to brother Weld's ultraism on the subject of marriage, he is quite mistaken if he fancies he has got far *ahead of us* in the human rights reform. We do *not* think his doctrine at all shocking: it is *altogether right.* But I am afraid I am too *proud* ever to exercise the right. The fact is we are living in such an artificial state of society that there are some feelings about which we dare not speak out, or act out the most natural & best feelings of our hearts. O! *when* shall we be "delivered from the *bondage of corruption* into the glorious liberty of the sons of God!" By the bye, it will be very important to establish this right, for the men of Mass[achusetts] stoutly declare that women who hold such sentiments of *equality* can never expect to be courted. They seem to hold out this as a kind of threat to deter us from asserting our rights, not *knowing whereunto this will grow.* But jesting is inconvenient says the Apostle: to business then.

Anti-slavery men are trying very hard to separate what God hath joined together. I fully believe that so far from keeping different moral reformations entirely distinct, that no such attempt can ever be successful. They are bound together in a circle, like the sciences; they blend with each other, like the colors of the rain bow; they are the parts only of our glorious whole, & that whole is Christianity, pure *practical* christianity. The fact is I believe — but don't be alarmed, for it is only I — that Men & Women will have to go out on their own responsibility, just like the prophets of old & declare the *whole* counsel of God to the people. The whole Church Government must come down; the clergy stand right in the way of reform, & I do not know but this stumbling block too must be removed *before* Slavery can be abolished, for the system is supported by *them;* it could not exist without the Church, as it is called. This grand principle must be mooted, discussed & established, viz. the Ministers of the Gospel are the successors of the *Prophets,* not of the *priests.* . . . The Church is built *not* upon the priests at all but upon the *prophets and apostles,* Jesus Christ being the chief corner stone.

This develops three important inferences; 1. True ministers are called, like Elisha from the plough & Amos from gathering sycamore fruit, Matthew from the receipt of custom & Peter and John from their fishing nets. 2. As prophets *never were paid,* so ministers ought not to be. 3. As there were *prophetesses* as well as prophets, so there *ought* to be now *female* as well as male ministers. Just let this one principle be established, & what will become of the power and sacredness of the pastoral office? Is brother Weld frightened at *my ultraism?* . . .

We never mention women's rights in our *lectures,* except so far as is necessary to urge them to meet their responsibilities. We speak of their *responsibilities* & leave *them* to *infer* their *rights*. I could cross this letter all over but must not encroach on your time.[1]

MAY THE LORD BLESS YOU MY DEAR BROTHERS
IS THE PRAYER OF YOUR SISTER IN JESUS,
A.E.G.

[1]To make their letters more compact, nineteenth-century correspondents, after they had filled one side of the paper, often turned the page sideways and wrote across what they had already written.

29

Resolutions Adopted by the Providence, Rhode Island, Ladies' Anti-Slavery Society

October 21, 1837

Published in The Liberator, *these resolutions show how the debates over women's rights were resounding within women's antislavery societies. Opposition to women's rights fueled the resolve of some societies.*

Whereas we believe the cause of the slave to be one of neglected humanity, and the southern portion of the Union to be one of the waste places of Zion, where for justice is oppression, and for righteousness, behold, a cry — therefore,

Resolved, That we act as moral agents and Christians fearlessly in this cause — thinking and acting in view of our accountability to our Maker — remembering that our rights are sacred and immutable, and founded on the liberty of the gospel, that great emancipation act for women. We further resolve, that we will not be turned aside from the object we have espoused, by the intimidations of ridicule, or the intoxicating flatteries of men and women, whose god is their selfishness, nor be cajoled into a selfish conceit of our superiority over the millions of females in our country, whose unuttered and unutterable cries of agony from oppression will, as they rise to heaven, shake terribly our guilty land; but we will turn our eyes, for example and imitation, to those philanthropists in Europe and America, who, through self-denial and perse-

"Voice of Woman," *The Liberator,* Nov. 3, 1837.

cution, have become pioneers in the cause of emancipation, some of whom we have seen face to face; and while they command our reverence, they call forth our gratitude as women for the shadowing out they have given of our rights, by means of the full light which their benevolent efforts have shed on the equality of the rights of man.

Whereas strenuous exertions are making at the present day, to counteract the disinterested labors of women in behalf of the oppressed, by representing them as "over-stepping the boundaries of their sex"—therefore,

Resolved, That we will not be influenced by such considerations, to shrink from the performance of the duty we owe to the suffering slave. Believing that woman can plead for the slave, without forsaking her "appropriate sphere of action," we rejoice that there are so many who possess strong minds and vigorous intellects, and are willing to labor in this cause. In accordance with these views, we deem the self-denying labors of the Misses Grimké worthy of all praise, and cordially approve of the course pursued by them in the cause of abolition.

Resolved, That we totally disapprove of the late Clerical Protests, regarding them as injudicious and unchristian; and believing that the Liberator has ever proven itself the firm and uncompromising friend of the slave, our confidence in the integrity and ability of its editor remain unshaken: Resolved, That the foregoing resolutions be forwarded to the editor of the Liberator.

SARAH PRATT, SECRETARY

30

PHILADELPHIA FEMALE ANTI-SLAVERY SOCIETY

Annual Report

1837

This report views the remarkable events of 1837 from the perspective of the largest women's antislavery society. The society's annual financial report shows how it spent its money, and its constitution probably served as the model for many other Garrisonian groups.

Annual Report, Philadelphia Female Anti-Slavery Society (Philadelphia: Merrihew & Thompson, 1838).

Allied as we are to the mighty phalanx of those who are contending for a "world's liberty," the retrospect of a year must furnish us many sources of encouragement, for we share with them a common lot; the success of one portion is the prosperity of the whole, and the perils and sufferings of a part are felt by all.

To overthrow an institution which has grown up, to giant size, in the heart of a mighty nation; which has its foundation in the strongest depraved principles of human nature; which is surrounded and sustained by the sanctions of law and public opinion, and protected by the suffrage of a false religion; to destroy and utterly lay waste such an institution, and to do it by moral influence on the minds of the community, is not the work of a day, or a year. Such a work is ours. It can be accomplished only by constant and unwearied effort, day after day, and year after year, by seizing every opportunity to pour a ray of light on the darkened understanding, or a softening influence on the hardened heart, till the mind of the nation is renovated, and the pillars of slavery are removed. . . .

On examining the records of our Society, we find that, during the past year, more than forty members have joined our ranks, an addition greatly exceeding the average increase of any of the preceding years of the Society's existence. . . .

A new measure for the advancement of our cause has this year been adopted by the Female Anti-Slavery Societies of this country, viz.: the holding of an "Anti-Slavery Convention of American Women." The object of this meeting was to afford the different associations an opportunity of conferring together respecting their modes of operation, and devising plans of united action; and also, that those women of America, whose souls are sickened by the oppressions that are done under the sun, might have an opportunity of together lifting up their voices, in remonstrance and entreaty, in behalf of their brethren and sisters in bonds, and of the eternal principles of justice.

A measure so novel, adopted by women, would, of course, excite surprise in many minds, and from some elicit censure. This we expected; for it we were prepared; nor have the editorial rebukes, sarcasm, and ridicule, which have been awarded us, exceeded our anticipation. To calm and manly argument we would have attentively listened, and respectfully replied; but to the coarse invective and rude jesting of which we were the subject, we deem it unfit to oppose sober reasoning, or even serious expostulation. . . .

The immediate results of our Convention have been given to the public, and it is needless to recapitulate them. A part of its fruits will probably be seen in the increased number of memorials which will be sent to

Congress during its present session. In pursuance of a plan proposed by the Convention, this Society, in the month of June, made arrangements with the Female Anti-Slavery Society of Pittsburg, to furnish every county in Pennsylvania with memorials [petitions] to Congress, praying for the abolition of slavery in the District of Columbia, and the Territories of the United States, and for the abolition of the slave trade between the states. In that portion of the State allotted to us, (judging from the reports of members to whom it was entrusted,) this work has been performed as thoroughly as our means of action permitted. The number of signatures obtained in Philadelphia exceeds that of the last year by about one thousand.

The benefits to the Anti-Slavery cause, arising from this department of our labor, are not to be calculated by the number of signatures appended to our memorials. We do not regard those visits as lost labor, where our request is denied, or that time wasted which is spent in unsuccessful efforts to convince persons of their duty to comply with it. Often, very often, the seed then laboriously sown, falls into good ground, and after a little season springs up, bringing forth fruit, some thirty, some sixty, some an hundred fold. Wherever our arguments find a lodgment in the mind, or our expostulations arouse the sympathies of the heart, there a victory (small though it may be) is won for the Anti-Slavery cause. And where no such result of our toil is perceptible, and we go from house to house, sad, and sick at heart with the selfishness, and ignorance, and pride with which we are repulsed, the reflection that we have suggested an important, and perhaps, new topic of thought and conversation, which will, no doubt, be pursued in some form or other, encourages us to persevere.

The prosperous state of our finances, which will be shown by the report of the treasurer, demands our grateful acknowledgments to Him who has liberally supplied us with the means of doing good, and should excite to the accomplishment of greater things in future. Our sale of useful and fancy articles, has been attended with unexpected success, and its pecuniary profits have considerably exceeded those of the sale of the preceding year. [See "Treasurer's Report" on page 141].

During the past year we have maintained a correspondence with several sister associations in Pennsylvania, and elsewhere. Their letters, informing us of their prosperity, detailing their plans of action, suggesting advice, or exhorting to constancy, have cheered and invigorated us. Intelligence respecting our beloved sisters, SARAH M. and ANGELINA E. GRIMKE, has also been frequently received. The year has been to them, one of trial and arduous labor, in which our hearts have sympathized. On reverting to the expression of our feelings, in regard to these dear coadjutors, contained in our last report, we find that many of our own anticipations respecting them have been realized, and although many have attributed the difficulties of their way, and the opposition which they meet, to the manner in which they have enlarged the sphere of labor at first proposed, in our opinion a large portion of it is the result of hostility to the truth, for which they are so earnestly contending. It is true, that respecting some of their measures, different opinions exist among abolitionists; but these measures have been adopted by themselves, and for them none are responsible but those who have chosen to become so, by endorsing them. Approval of the course which they have pursued has already been expressed by a majority of this Society, and published in the "National Enquirer," and the "Liberator." . . .

Constitution of the Philadelphia Female Anti-Slavery Society.

WHEREAS, more than two millions of our fellow countrymen [*sic*], of these United States, are held in abject bondage; and whereas, we believe that slavery and prejudice against color are contrary to the laws of God, and to the principles of the our far-famed Declaration of Independence, and recognising [*sic*] the right of the slave to immediate emancipation; we deem it our duty to manifest our abhorrence of the flagrant injustice and deep sin of slavery, by united and vigorous exertions for its speedy removal, and for the restoration of the people of color to their inalienable rights. For these purposes, we, the undersigned, agree to associate ourselves under the name of the "THE PHILADELPHIA FEMALE ANTI-SLAVERY SOCIETY."

ARTICLE I.

The object of this society shall be to collect and disseminate correct information of the character of slavery, and of the actual condition of the slaves and free people of color, for the purpose of inducing the community to adopt such measures, as may be in their power, to dispel the prejudice against people of color, to improve their condition, and to bring about the speedy abolition of slavery.

ARTICLE II.

Any female uniting in these views, and contributing to the funds, shall be a member of the Society.

ARTICLE III.

The officers of the Society shall be a President, a Vice President, a Recording Secretary, a Corresponding Secretary, a Treasurer, and Librarian, who, with six other members, shall constitute a Board of Managers, to whom shall be intrusted [*sic*] the business of the Society, and the management of its funds. They shall keep a record of their proceedings, which shall be laid before the Society, at each stated meeting. They shall have power to fill any vacancy that may occur in their number, till the next annual meeting.

ARTICLE IV.

The President shall preside at all meetings of the Society, and shall have power to call special meetings of the Society and of the Board.

ARTICLE V.

The Vice President shall perform the duties of the President in her absence.

ARTICLE VI.

The Recording Secretary shall keep a record of the transactions of the Society, and notify all meetings of the Society.

ARTICLE VII.

The Corresponding Secretary shall keep all communications addressed to the Society, and manage all the correspondence with any other bodies or individuals, according to the directions of the Society or of the Managers.

ARTICLE VIII.

The Treasurer shall collect the subscriptions and grants to the Society, make payments according to its directions, and those of its Managers, and present an audited report at each annual meeting.

ARTICLE IX.

The Librarian shall take charge of all books and pamphlets belonging to the Society, and conform to the rules prescribed by the Society, for the management of the library.

ARTICLE X.

The Managers shall meet once a month, or oftener if necessary, on a day fixed by themselves, and stated meetings of the Society shall be held on the second Fifth-day in every month.

ARTICLE XI.

The annual meeting shall be held on the second Fifth-day (Thursday) of the First month (January) at which time the reports of the Board and Treasurer shall be presented, and the officers for the ensuing year elected.

ARTICLE XII.

It is especially recommended that the members of this Society should entirely abstain from purchasing the products of slave labor, that we may be able consistently to plead the cause of our brethren in bonds.

ARTICLE XIII.

This Constitution may be altered at any stated meeting, by the vote of two-thirds of the members present, notice having been given at a previous meeting.

OFFICERS FOR THE ENSUING YEAR. PRESIDENT, SARAH PUGH. VICE PRESIDENT, ANNA M. HOPPER. RECORDING SECRETARY, SARAH M. DOUGLASS. CORRESPONDING SECRETARY, MARY GREW. TREASURER, MARGARETTA FORTEN. MANAGERS: Lucretia Mott, Lydia White, Sarah Forten, Mary Needles, Grace Douglass, Susan Haydock. The meetings of the Society are held on the second Thursday of every month, at Sandiford Hall, Haines Street, between Sixth and Seventh, Arch and Race Streets.

TREASURER'S REPORT.

The Philadelphia Female Anti-Slavery Society, in Account with M. Forten, Treasurer.

DR.

Date	Item	$	cts.
1837. Jan.	To cash paid for Pennsylvania Hall stock, for five copies of the Liberator,	$400 10	
Feb.	for rent of Temperance Hall for meetings,	12	50
	for printing Annual Report,	12	53
March	for rent of Adelphi hall for meeting,	10	50
April	do. Temperance Hall do.	6	
May	to A.S. Convention of Women, on account of our pledge,	25	
June	to City and County A.S. Society, on account of our pledge,	70	
	for printing memorials,	18	
Nov.	to Messrs. Burleigh and Gunn, for the expenses of their voyage,	50	
Dec.	to Junior A.S. Society,	20	
	for Facts for the People and Alton Observer,	9	
	for lighting and cleaning Sandiford Hall,	1	75
	to M. Moore for articles furnished for last year's sale,	1	
1838.	for postage,	4	
Jan.	to City and County A.S. Society, on account of our pledge,	60	
	to balance in the Treasury,	332	23
		1042	51

CR.

Date	Item	$	cts.
1837. Jan.	By balance in the treasury,	$114	81
	annual subscriptions,	85	23
Feb.	collection at public meeting,	27	2
April	collections made for purchasing Pennsylvania Hall stock,	203	
	collection at public meeting,	17	28
	annual subscriptions,	36	35
1838.	Subscription for A.S. Record,	2	
Jan.	Cash received from managers of the A.S. sale,	459	18
	contributions and donations during the year,	26	83
	sale of Anti-Slavery tracts,		78
		1042	51[1]

[1]Figures in credit column actually total $972.48, indicating an error in the original accounts of $70.03.

31

ANGELINA GRIMKÉ

"Human Rights Not Founded on Sex": Letter to Catharine Beecher

August 2, 1837

Answering Beecher's Essay on Slavery and Abolitionism, *which appeared in March 1837, Angelina in the summer of 1837 composed a series of letters that defended the activism of antislavery women. Her twelfth letter addressed the issue of women's rights. Published individually in* The Liberator, *her letters appeared in book form in 1838.*

DEAR FRIEND:

Since I engaged in the investigation of the rights of the slave, I have necessarily been led to a better understanding of my own; for I have found the Anti-Slavery cause to be the high school of morals in our land — the school in which human rights are more fully investigated, and better understood and taught, than in any other benevolent enterprise. Here one great fundamental principle is disinterred, which, as soon as it is uplifted to public view, leads the mind into a thousand different ramifications, into which the rays of this central light are streaming with brightness and glory. Here we are led to examine why human beings have any rights. It is because they are moral beings; the rights of all men, from the king to the slave, are built upon their moral nature: and as all men have this moral nature, so all men have essentially the same rights. These rights may be plundered from the slave, but they cannot be alienated: his right and title to himself is as perfect now, as is that of Lyman Beecher: they are written in his moral being, and must remain unimpaired as long as that being continues. Now it naturally occurred to me, that if rights were founded in moral being, then the circumstance of sex could not give to man higher rights and responsibilities, than to woman. To suppose that it did, would be to deny the self-evident truth, "that the physical constitution is the mere instrument of the moral na-

---

The Liberator, Aug. 2, 1837; reprinted in Angelina E. Grimké, *Letters to Catherine E. Beecher, in Reply to an Essay on Slavery and Abolitionism* (Boston: Knapp, 1838).

ture." To suppose that it did, would be to break up utterly the relations of the two natures, and to reverse their functions, exalting the animal nature into a monarch, and humbling the moral into a slave; "making the former a proprietor, and the latter its property." When I look at human beings as moral beings, all distinction in sex sinks to insignificance and nothingness; for I believe it regulates rights and responsibilities no more than the color of the skin or the eyes. My doctrine then is, that whatever it is morally right for man to do, it is morally right for woman to do. Our duties are governed, not by difference of sex, but by the diversity of our relative connections in life, and the variety of gifts and talents committed to our care, and the different eras in which we live.

This regulation of duty by the mere circumstance of sex, rather than by the fundamental principle of moral being, has led to all that multifarious train of evils flowing out of the anti-christian doctrine of masculine and feminine virtues. By this doctrine, man has been converted into the warrior, and clothed in sternness, and those other kindred qualities, which, in the eyes of many, belong to his character as a man; whilst woman has been taught to lean upon an arm of flesh, to sit as a soul arrayed "in gold and pearls, and costly array," to be admired for her personal charms, and caressed and humored like a spoiled child, or converted into a mere drudge to suit the convenience of her lord and master. This principle has spread desolation over the whole moral world, and brought into all the diversified relations of life, "confusion and every evil work." It has given to man a charter for the exercise of tyranny and selfishness, pride and arrogance, lust and brutal violence. It has robbed woman of essential rights, the right to think and speak and act on all great moral questions, just as men think and speak and act; the right to share their responsibilities, dangers, and toils; the right to fulfill the great end of her being, as a help meet for man, as a moral, intellectual and immortal creature, and of glorifying God in her body and her spirit which are His. Hitherto, instead of being a help meet to man, in the highest, noblest sense of the term, as a companion, a co-worker, an equal; she has been a mere appendage of his being, and instrument of his convenience and pleasure, the pretty toy, with which he wiled away his leisure moments, or the pet animal whom he humored into playfulness and submission. Woman, instead of being regarded as the equal of man, has uniformly been looked down upon as his inferior, a mere gift to fill up the measure of his happiness. In the poetry of "romantic gallantry," it is true, she has been called the "last best gift of God to man;" but I believe I speak forth the words of truth and soberness when I affirm, that woman never was given to man. She was created, like him, in the image of God, and crowned with glory and honor; created only a little lower than the angels,—not, as is too generally presumed, a little lower than man; on her brow, as well as on his, was placed the "diadem of beauty," and in her hand the scepter of universal dominion. . . . Let us examine the account of her creation. "And the rib which the Lord God had taken from man, made he a woman, and brought her unto the man." Not as a gift—for Adam immediately recognized her as a part of himself—(this is now "bone of my bone, and flesh of my flesh")—a companion and equal, not one hair's breadth beneath him in the greatness of her moral being; not one iota subject to him, for they both stood on the same platform of human rights, immediately under the government of God only. This idea of woman's being "the last best gift of God to man," however pretty it may sound to the ears of those who love to discourse upon the poetry of "romantic gallantry, and the generous promptings of chivalry," has nevertheless been the means of sinking her from an end into a mere means—of turning her into an appendage to, instead of recognizing her as part of man—of destroying her individuality, and rights, and responsibilities, and merging her moral being into that of man. Instead of Jehovah being her king, her lawgiver, and her judge, she has been taken out of the exalted scale of existence in which He placed her, and crushed down under the feet of man. . . .

Measure her rights and duties by the sure, unerring standard of moral being, not by the false rights and measures of a mere circumstance of her human existence, and then will it become a self-evident truth, that whatever it is morally right for a man to do, it is morally right for a woman to do. I recognize no rights but human rights—I know nothing of men's rights and women's rights; for in Christ Jesus, there is neither male or female; and it is my solemn conviction, that, until this important principle of equality is recognized and carried out into practice, that vain will be the efforts of the church to do anything effectual for the permanent reformation of the world. Woman was the first transgressor, and the first victim of power. In all the heathen nations, she has been the slave of man, and no Christian nation has ever acknowledged her rights. Nay more, no Christian Society has ever done so either, on the broad and solid basis of humanity. I know that in some few denominations, she is permitted to preach the gospel; but this is not done from a conviction of her equality as a human being, but of her equality in spiritual gifts—for we find that woman, even in these Societies, is not allowed to make the Discipline by which she is to be governed. Now, I believe it is her right to be consulted in all the laws and regulations by which she is to be governed, whether in Church or State, and that the present arrangement of

Society, on those points, are a violation of human rights, an usurpation of power over her, which is working mischief, great mischief, in the world. If Ecclesiastical and Civil governments are ordained of God, then I contend that woman has just as much right to sit in solemn counsel in Conventions, Conferences, Associations, and General Assemblies, as man — just as much right to sit upon the throne of England, or in the Presidential chair of the United States, as man. . . .

I believe the discussion of Human Rights at the North has already been of immense advantage to this country. It is producing the happiest influence upon the minds and hearts of those who are engaged in it; . . . Indeed, the very agitation of the question, which it involved, has been highly important. Never was the heart of man so expanded; never were its generous sympathies so generally and so perseveringly excited. These sympathies, thus called into existence, have been useful preservatives of national virtue. I therefore do wish very much to promote the Anti-Slavery excitement at the North, because I believe it will prove a useful preservative of national virtue. . . .

The discussion of the wrongs of slavery has opened the way for the discussion of other rights, and the ultimate result will most certainly be "the breaking of every yoke," the letting the oppressed of every grade and description go free — an emancipation far more glorious than any the world has ever yet seen, an introduction into that liberty wherewith Christ hath made his people free. . . .

THY FRIEND,
ANGELINA E. GRIMKÉ

32

SARAH GRIMKÉ

"Legal Disabilities of Women": Letter to Mary Parker

September 6, 1837

Addressed to Mary Parker, President of the Boston Female Anti-Slavery Society, Sarah's Letters ***were primarily devoted to providing a scriptural basis***

The Liberator, Feb. 2, 1838; reprinted in Sarah Grimké, ***Letters on the Equality of the Sexes and the Condition of Woman*** (Boston: Knapp, 1838).

for women's equality. This twelfth letter, however, offered an effective critique of secular and ecclesiastical laws that kept women inferior, just as laws kept slaves in bondage. Published individually in The Liberator, *the* Letters *appeared as a book in 1838.*

MY DEAR SISTER:
There are few things which present greater obstacles to the improvement and elevation of woman to her appropriate sphere of usefulness and duty, than the laws which have been enacted to destroy her independence, and crush her individuality; laws which, although they are framed for her government, she has had no voice in establishing, and which rob her of some of her essential rights. Woman has no political existence. With the single exception of presenting a petition to the legislative body, she is a cipher in the nation; or, if not actually so in representative governments, she is only counted, like the slaves of the South, to swell the number of law-makers who form decrees for her government, with little reference to her benefit, except so far as her good may promote their own. . . . I shall confine myself to the laws of our country. These laws bear with peculiar rigor on married women. Blackstone, in the chapter entitled "Of husband and wife," says:—

> By marriage, the husband and wife are one person in law; that is, the very being, or legal existence of the woman is suspended during the marriage, or at least is incorporated and consolidated into that of the husband under whose wing, protection and cover she performs everything.[1]
>
> For this reason, a man cannot grant anything to his wife, or enter into covenant with her; for the grant would be to suppose her separate existence, and to covenant with her would be to covenant with himself; and therefore it is also generally true, that all compacts made between husband and wife when single, are voided by the intermarriage. A woman indeed may be attorney for her husband, but that implies no separation from, but rather a representation of, her love.

Here now, the very being of a woman, like that of a slave, is absorbed in her master. All contracts made with her, like those made with slaves by their owners, are a mere nullity. Our kind defenders have legislated away almost all our legal rights, and in the true spirit of such injustice and oppression, have kept us in ignorance of those very laws by which we

[1] In his *Commentaries on the Laws of England* (1765–69), William Blackstone (1723–1780) interpreted English Common Law traditions, including *feme covert* laws through which the legal identity of married women was absorbed into that of their husbands.

are governed. They have persuaded us, that we have no right to investigate the laws, and that, if we did, we could not comprehend them; they alone are capable of understanding the mysteries of Blackstone, &c. . . .

> The husband is bound to provide his wife with necessaries by law, as much as to himself; and if she contracts debts for them, he is obliged to pay for them; but for anything besides necessaries, he is not chargeable.

Yet a man may spend the property he has acquired by marriage at the ale-house, the gambling table, or in any other way that he pleases. Many instances of this kind have come to my knowledge; and women, who have brought their husbands handsome fortunes, have been left, in consequence of the wasteful and dissolute habits of their husbands; in straitened circumstances, and compelled to toil for the support of their families. . . .

> If the wife be injured in her person or property, she can bring no action for redress without her husband's concurrence, and his name as well as her own; neither can she be sued, without making her husband a defendant.

This last that "a wife can bring no action," &c. is similar to the law respecting slaves. A slave cannot bring a suit against his master, or any other person, for an injury — his master must bring it. So if any damages are recovered for an injury committed on a wife, the husband pockets it; in the case of the slave, the master does the same.

In criminal prosecutions, the wife may be indicted and punished separately, unless there be evidence of coercion from the fact that the offence was committed in the presence, or by the command of her husband. "A wife is excused from punishment for theft committed in the presence, or by the command of her husband."

It would be difficult to frame a law better calculated to destroy the responsibility of woman as a moral being, or a free-agent. Her husband is supposed to possess unlimited control over her; and if she can offer the flimsy excuse that he bade her to steal, she may break the eighth commandment with impunity, as far as human laws are concerned.

> Our law, in general, considers man and wife as one person; yet there are some instances in which she is separately considered, as inferior to him and acting by his compulsion. Therefore, all deeds executed, and acts done by her during her coverture (i.e. marriage) are void, except it be a fine, or like matter of record, in which case she must be solely and secretly examined, to learn if her act be voluntary.

Such a law speaks volumes of the abuse of that power which men have vested in their own hands. Still the private examination of a wife, to know whether she accedes to the disposition of property made by her husband is, in most cases, a mere form; a wife dares not do what will be disagreeable to one who is, in his own estimation, her superior, and who makes her feel, in the privacy of domestic life, that she has thwarted him. . . .

> The husband, by the old law, might give his wife moderate correction, as he is to answer for her misbehavior. The law thought it reasonable to entrust him with this power of retraining her by domestic chastisement. The courts of law will still permit a husband to restrain a wife of her liberty, in case of any gross misbehavior.

What a mortifying proof this law affords, of the estimation in which woman is held! She is placed completely in the hands of a being subject like herself to the outbursts of passion, and therefore unworthy to be trusted with power. Perhaps I may be told respecting this law, that it is a dead letter, as I am sometimes told about the slave laws; but this is not true in either case. The slaveholder does kill his slave by moderate correction, as the law allows; and many a husband, among the poor, exercises the right given to him by the law, of degrading woman by personal chastisement. And among the higher ranks, if actual imprisonment is not resorted to, women are not infrequently restrained of the liberty of going to places of worship by irreligious husbands, and of doing many other things about which, as moral and responsible beings, they should be the sole judges.

Such laws remind me of the reply of some little girls at a children's meeting held recently at Ipswich. The lecturer told them that God had created four orders of beings with which he had made us acquainted through the Bible. The first was angels, the second was man, the third beasts; and now, children, what is the fourth? After a pause, several girls replied, "WOMEN."

> A woman's personal property by marriage becomes absolutely her husband's, which, at his death, he may leave entirely away from her.

And farther, all the avails of her labor are absolutely in the power of her husband. All that she acquires by her industry is his; so that she cannot, with her own honest earnings, become the legal purchaser of any property. If she expends her money for articles of furniture, to contribute to the comfort of her family, they are liable to be seized for her husband's debts: and I know an instance of a woman, who by labor and economy had scraped together a little maintenance for herself and a do-little husband, who was left, at his death, by virtue of his last will and testament, to be supported by charity. . . .

The laws above cited are not very unlike the slave laws of Louisiana. "All that a slave possesses belongs to his master; he possesses nothing of his own, except what his master chooses he should possess." . . .

As these abuses do exist, and women suffer intensely from them, our brethren are called upon in this enlightened age, by every sentiment of honor, religion, and justice, to repeal these unjust and unequal laws, and restore to woman those rights which they have wrested from her. Such laws approximate too nearly to the laws enacted by slaveholders for the government of their slaves, and must tend to debase and depress the mind of that being whom God created as a help meet for man, or helper "like unto himself," and designed to be his equal and his companion. Until such laws are annulled, woman never can occupy that exalted station for which she was intended by her Maker. And just in proportion as they are practically disregarded, which is the case to some extent, just so far is woman assuming that independence and nobility of character which she ought to exhibit. . . .

Ecclesiastical bodies, I believe, without exception, follow the example of legislative assemblies, in excluding woman from any participation in forming the discipline by which she is governed. The men frame the laws, and, with few exceptions, claim to execute them on both sexes. In ecclesiastical, as well as civil courts, woman is tried and condemned, not by a jury of her peers, but by beings, who regard themselves as her superiors in the scale of creation. Although looked upon as an inferior, when considered as an intellectual being, woman is punished with the same severity as man, when she is guilty of moral offenses. Her condition resembles, in some measure, that of the slave, who, while he is denied the advantages of his more enlightened master, is treated with even greater rigor of the law. Hoping that in the various reformations of the day, women may be relieved from some of their legal disabilities, I remain,

THINE IN THE BONDS OF WOMANHOOD,
SARAH M. GRIMKÉ

33

SARAH GRIMKÉ

"Relation of Husband and Wife": Letter to Mary Parker

September 1837

One of the most important aspects of the sisters' vision of women's rights was their hope to transform private marital relations as well as women's place in public life. This letter addressed that hope.

My Dear Sister:
Perhaps some persons may wonder that I should attempt to throw out my views on the important subject of marriage, and may conclude that I am altogether disqualified for the task, because I lack experience. However, I shall not undertake to settle the specific duties of husbands and wives, but only to exhibit opinions based on the word of God, and formed from a little knowledge of human nature, and close observation of the working of generally received notions respecting the dominion of man over woman. . . .

I have already shown, that man has exercised the most unlimited and brutal power over woman, in the peculiar character of husband,—a word in most countries synonymous with tyrant. . . .

[T]hat state which was designed by God to increase the happiness of woman as well as man, often proves the means of lessening her comfort, and degrading her into the mere machine of another's convenience and pleasure. Woman, instead of being elevated by her union with man, which might be expected from an alliance with a superior being, is in reality lowered. She generally loses her individuality, her independent character, her moral being. . . .

In the wealthy classes of society, and those who are in comfortable circumstances, women are exempt from great corporeal exertion, and are protected by public opinion, and by the genial influence of Christianity, from much physical ill treatment. Still, there is a vast amount of secret suffering endured, from the forced submission of women to the opinions and whims of their husbands. Hence they are frequently driven

---

The Liberator, Feb. 2, 1838; reprinted in Grimké, *Letters on the Equality of the Sexes.*

to use deception, to compass their ends. They are early taught that to appear to yield, is the only way to govern. . . .

[W]omen, among the lowest classes of society, as far as my observation has extended, suffer intensely from the brutality of their husbands. Duty as well as inclination has led me, for many years, into the abodes of poverty and sorrow, and I have been amazed at the treatment which women receive at the hands of those who arrogate to themselves the epithet of protectors. Brute force, the law of violence, rules to a great extent in the poor man's domicil; and woman is little more than his drudge. They are less under the supervision of public opinion, less under the restraints of education, and unaided or unbiased by the refinements of polished society. Religion, wherever it exists, supplies the place of all these; but the real cause of woman's degradation and suffering in married life is to be found in the erroneous notion of her inferiority to man; and never will she be rightly regarded by herself, or others, until this opinion, so derogatory to the wisdom and mercy of God, is exploded, and woman arises in all the majesty of her womanhood, to claim those rights which are inseparable from her existence as an immortal, intelligent, and responsible being. . . .

But it is strenuously urged by those, who are anxious to maintain their usurped authority, that wives are, in various passages of the New Testament, commanded to obey their husbands. Let us examine these texts.

Eph. 5:22. "Wives, submit yourselves unto your own husbands as unto the Lord. As the church is subject unto Christ, so let the wives be to their own husbands in every thing."

Col. 3:18. "Wives, submit yourselves unto your own husband, as it is fit in the Lord."

1st Pet. 3:2. "Likewise ye wives, be in subjection to your own husband; that if any obey not the word, they may also without the word be won by conversation of the wives."

Accompanying all these directions to wives, are commands to husbands.

Eph. 5:25. "Husbands, love your wives even as Christ loved the Church, and gave himself for it. So ought men to love their wives as their own bodies. He that loveth his wife, loveth himself."

Col. 3:19. "Husbands, love your wives, and be not bitter against them."

1st Pet. 3:7. "Likewise ye husbands, dwell with them according to knowledge, giving honor unto the wife as unto the weaker vessel, and as being heirs together of the grace of life."

I may just remark, in relation to the expression "weaker vessel," that the word in the original has no reference to intellect: it refers to physical weakness merely.

The apostles were writing to Christian converts, and laying down rules for their conduct towards their unconverted consorts. It no doubt frequently happened, that a husband or a wife would embrace Christianity, while their companions clung to heathenism, and husbands might be tempted to dislike and despise those, who pertinaciously adhered to their pagan superstitions. And wives who, when they were pagans, submitted as a matter of course to their heathen husbands, might be tempted knowing that they were superior as moral and religious characters, to assert that superiority, by paying less deference to them than heretofore. . . .

From an attentive consideration of these passages, and of those in which the same words "submit," "subjection," are used, I cannot but believe that the apostles designed to recommend to wives, as they did to subjects and to servants, to carry out the holy principle laid down by Jesus Christ, "Resist not evil." And this without in the least acknowledging the right of the governors, masters, or husbands, to exercise the authority they claimed. The recognition of the existence of evils does not involve approbation of them.

THINE IN THE BONDS OF WOMANHOOD,
SARAH M. GRIMKÉ

THE ANTISLAVERY MOVEMENT SPLITS OVER THE WOMEN'S RIGHTS QUESTION, 1837–1840

34

ANGELINA GRIMKÉ WELD

Speech at Pennsylvania Hall

Philadelphia, May 16, 1838

In her last public speech Angelina Grimké proceeded despite the siege of the building by an anti-abolitionist mob. This was her only speech to be transcribed — in a shorthand process that was called phonographic. Her oratory shows that she had become adept at improvisation. Bracketed comments about the mob's assault were added by the transcriber.

Men, brethren, and fathers-mothers, daughters and sisters, what came ye to see? A reed shaken with the wind? Is it curiosity merely, or a deep sympathy with the perishing slave, that has brought this large audience together? [A yell from the mob without the building.] Those voices without ought to awaken and call out our warmest sympathies. Deluded Beings! "they know not what they do."[1] They know not that they are undermining their own rights and their own happiness, temporal and eternal. Do you ask, "what has the North to do with slavery?" Hear it — hear it. Those voices without tell us that the spirit of slavery is *here,* and has been roused to wrath by our abolition speeches and conventions: for surely liberty would not foam and tear herself with rage, because her friends are multiplied daily, and meetings are held in quick succession to set forth her virtues and extend her peaceful kingdom. This opposition shows that slavery has done its deadliest work in the hearts of our citizens.

Do you ask, then, "what has the North to do?" I answer, cast out first the spirit of slavery from your own hearts, and then lend your aid to con-

[1] Luke 23:34.

Samuel Webb, ed., *History of Pennsylvania Hall, Which Was Destroyed by a Mob, on the 17th of May, 1838* (Philadelphia: Merrihew and Gunn, 1838), 123–26.

vert the South. Each one present has a work to do, be his or her situation what it may, however limited their means, or insignificant their supposed influence. The great men of this country will not do this work; the church will never do it. A desire to please the world, to keep the favor of all the parties and of all conditions, makes them dumb on this and every other unpopular subject. They have become worldly-wise, and therefore God, in his wisdom, employs them not to carry on his plans of reformation and salvation. He hath chosen the foolish things of the world to confound the wise, and the weak to overcome the mighty.[2]

As a Southerner I feel that it is my duty to stand up here tonight and bear testimony against slavery. I have seen it—I have seen it. I know it has horrors that can never be described. I was brought up under its wing: I witnessed for many years its demoralizing influences, and its destructiveness to human happiness. It is admitted by some that the slave is not happy under the *worst* forms of slavery. But I have *never* seen a happy slave. . . .

[Just then stones were thrown at the windows,—a great noise without, and commotion within.] What is a mob? What would the levelling of this Hall be? Any evidence that we are wrong or that slavery is a good and wholesome institution? What if the mob should now burst in upon us, break up our meeting and commit violence upon our persons—would this be anything compared with what the slaves endure? No, no: and we do not remember them "as bound with them," if we shrink in time of peril, or feel unwilling to sacrifice ourselves, if need be, for their sake. [Great noise.] . . .

Much will have been done for the destruction of Southern slavery when we have so reformed the North that no one here will be willing to risk his reputation by advocating or even excusing the holding of men as property. The South know it, and acknowledged that as fast as our principles prevail, the hold of the master must be relaxed. [Another outbreak of mobocratic spirit, and some confusion in the house.] . . .

I feel that all this disturbance is but an evidence that our efforts are the best that could have been adopted, or else friends of slavery would not care for what we say and do. The South know what we do. I am thankful that they are reached by our efforts. Many times have I wept in the land of my birth over the system of slavery. I knew none who sympathized in my feelings—I was unaware that any efforts as these were being made to deliver the oppressed—no voice in the wilderness was heard calling on the people to repent and do works meet for repen-

[2] I Corinthians 1:27–29.

tance[3]—and my heart sickened within me. Oh, how should I have rejoiced to know that such efforts as these were being made. I only wonder that I had such feelings. I wonder when I reflect under what influence I was brought up, that my heart is not harder than the nether millstone. But in the midst of temptation, I was preserved and my sympathy grew warmer, and my hatred of slavery more inveterate, until at last I have exiled from my native land because I could no longer endure to hear the wailing of the slave. I fled to the land of Penn; for here, thought I, sympathy for the slave will surely be found. But I found it not. The people were kind and hospitable, but the slave had no place in their thoughts. Whenever questions were put to me as to his condition, I felt that they were dictated by an idle curiosity, rather than by that deep feeling which would lead to effort for his rescue. I therefore shut up my grief in my own heart. I remembered that I was a Carolinian, from a state which framed this iniquity by law. I knew that throughout her territory was continued suffering, on the one part, and continual brutality and sin on the other. Every Southern breeze wafted to me the discordant tones of weeping and wailing, shrieks and groans, mingled with prayers and blasphemous curses. I thought there was no hope; that the wicked would go on in his wickedness, until he had destroyed both himself and his country. My heart sunk within me at the abominations in the midst of which I had been born and educated. What will it avail, cried I in bitterness of spirit, to expose to the gaze of strangers the horrors and pollutions of slavery, when there is no ear to hear nor heart to feel and pray for the slave. . . . But how different do I feel now! Animated with hope, Nay, with an assurance of the triumph of liberty and good will to man, I will lift up my voice like a trumpet, and show this people their transgression, their sins of omission towards the slave, and what they can do towards affecting Southern mind[s], and overthrowing Southern oppression. . . . [Shoutings, stones thrown against the windows, &c.]

There is nothing to be feared from those who would stop our mouths, but they themselves should fear and tremble. The current is even now setting fast against them. . . . [Mob again disturbed the meeting.]

We often hear the question asked, "What shall we do?" Here is an opportunity for doing something now. Every man and woman present may do something by showing that we fear not a mob, and, in the midst of threatening and revilings, by opening our mouths for the dumb and pleading the cause of those who are ready to perish.

[3] Matthew 3:3.

To work as we should in this cause, we must know what Slavery is. Let me urge you then to buy the books which have been written on this subject and read them, and then lend them to your neighbors. Give your money no longer for things which pander to pride and lust, but aid in scattering "the living coals of truth" upon the naked heart of this nation,—in circulating appeals to the sympathies of Christians in behalf of the outraged and suffering slave. . . .

Women of Philadelphia! allow me as a Southern woman, with much attachment to the land of my birth, to entreat you to come up to this work. Especially let me urge you to petition. Men may settle this and other questions at the ballot-box, but you have no such right; it is only through petitions that you can reach the Legislature. It is therefore peculiarly your duty to petition. Do you say, "It does no good?" The South already turns pale at the number sent. They have read the reports of the proceedings of Congress, and there have seen that among the other petitions were very many from the women of the North on the subject of slavery. This fact has called the attention of the South to the subject. How could we expect to have done more as yet? Men who hold the rod over slaves, rule in the councils of the nation: and they deny our right to petition and to remonstrate against abuses of our sex and of our kind. We have these rights, however, from our God. Only let us exercise them: and though often turned away unanswered, let us remember the influence of importunity upon the unjust judge, and act accordingly. The fact that the South look with jealousy upon our measures shows that they are effectual. There is, therefore, no cause for doubting or despair, but rather for rejoicing.

It was remarked in England that women did much to abolish Slavery in her colonies. . . . When the women of these States send up to Congress such as petition, our legislators will arise as did those of England, and say, "When all the maids and matrons of the land are knocking at our doors we must legislate." Let the zeal and love, the faith and works of our English sisters quicken ours—that while the slaves continue to suffer, and when they shout deliverance, we may feel that satisfaction of *having done what we could.*

35

HENRY CLARKE WRIGHT

Letter to The Liberator

New York, May 15, 1840

Although Wright was biased against the "new organization," his account of their secession from the American Anti-Slavery Society quoted their leaders correctly, and accurately represented their stated motives.

MY BROTHER:

The deed is done. The spirit of new organization, which is the spirit of slavery in disguise,—has, to the delight of pro-slavery ministers, editors and politicians, effected a division in the American Anti-Slavery Society. The Journal of Commerce, the Courier and Enquirer, the Commercial Advertiser and Herald, well known and staunch advocates of slavery and enemies of human rights, have extended to those who have gone out from among us the right hand of fellowship. A new society is formed, called the American and Foreign Anti-Slavery Society.

What is the basis of this new organization? THE AMERICAN SOCIETY REFUSED TO PRECLUDE WOMEN FROM THE RIGHTS OF MEMBERSHIP. This, and this only, according to the public statement of those who formed it, is the cause. Women have been invited to join the society from the beginning. Women have joined, under the impression that, according to the Constitution, they should be entitled to all the privileges of membership. Last year, urged on to the work by Sarah and Angelina Grimké, women were present at the anniversary, spoke and voted in the meetings, and were appointed on committees by Gerrit Smith. This year, women came forward to join with their brethren in this labor of love to sustain the cause of human rights. A business committee was appointed, consisting of Lewis Tappan, Charles W. Denison, Amos A. Phelps, Abby Kelley and others. Because a woman was on the committee, Lewis Tappan declined serving — giving as his reasons the following —

1. "To put a woman on a committee with men, is contrary to the Constitution of the Society."

"The New National Organization," *The Liberator,* May 22, 1840.

2. "It is throwing a fire brand into the anti-slavery ranks."
3. "It is contrary to the usages of civilized society." A. A. Phelps and C. W. Denison, refused for the same reasons, adding this further reason —
4. "It was contrary to the gospel and to their consciences."

Rev. Messrs. Denison and Dunbar, though they were conscientiously opposed to having women enjoy all the privileges of membership, strenuously urged the women to vote against A. Kelley's being on the committee. When they could not succeed in excluding members from the constitutional rights of membership, Lewis Tappan and C. W. Denison arose in the meeting and gave the following notice: "All who voted against the appointment of women on committees are requested to meet and form an American and Foreign Anti-Slavery Society."

Those who were opposed to women's acting in our Anti-Slavery Society with men, on the principle of equality of rights, went off and formed a new society.

Thus, because the society would not deny to the women their constitutional rights on the anti-slavery platform, Lewis Tappan . . . and others have seceded and set up another organization, from which woman is to be virtually excluded, because she is woman.

With what propriety is this "new organization" called an Anti-Slavery society? Anti-Slavery asserts the equality of human rights. It looks at man as man; it estimates man, not by his accidents and adjuncts — not by sex, color, tribe, nation, country or condition — but as man — and recognizes every human being, without regard to complexion, sex, or condition, as being the image and representative of God on earth — having an equal dominion over this lower world. But here is a society which excludes one half of the human race from its platform because of sex. How can any woman who feels for the virtue, honor and true dignity and glory of her sex, give any countenance to such a pro-slavery organization? The only reason given by Lewis Tappan and C. W. Denison, when they invited us to aid in its formation, was, in substance, to have a society which should exclude women, because they are women, from participating in its business transactions — from speaking, voting, and acting on committees in business meetings — for they invited only those to attend who were opposed to women thus acting. Women may speak, vote and act on committees, but by themselves. They must not presume to speak, vote and act with men in Anti-Slavery meetings.

What spirit is this but the spirit of despotism, which crushes women on three-fourths of the globe? The same spirit that originated the American and Foreign Anti-Slavery Society, has precluded woman from the society of man all over the East — shut her up in harems and seraglios, to keep her from mingling with her brother in the affairs of life — which has ever made woman the slave, the uncomplaining, suffering slave of man. Because her God made her a woman, she is thrust off from that platform where her brother acts. Where are those women, Sarah M. Grimké and Angelina E. Weld, who, for a brief space, so powerfully advocated the cause of human rights with tongue and pen? Has God stricken them with paralysis? Is their light gone out in total darkness? Oh that God would move them to speak once more in this crisis.

That same spirit which excludes the colored man, because of his color, from our schools, colleges, churches, legislatures, travelling conveyances, and from our social sympathies and circles, has originated in this new society. The colored people may go off by themselves to Liberia or elsewhere, and act by themselves; but they must not mingle with the whites. So has this new organization colonized the women. "The women may speak, vote, and act by themselves, but not with us, They are women." . . . And this is the end of Lewis Tappan's regard to the holy principles of human rights! He will no longer work in the American Anti-Slavery Society to sustain and propagate these eternal, immutable principles of liberty, justice, and equality, solely because that society will not deny to woman her right as a human being to speak and act with her brethren on the anti-slavery platform. How art thou fallen, my brother! When thou art interrogated why thou didst lead in this unhappy division, at heaven's tribunal, wilt thou answer — Woman — WOMAN — was the cause? She came forward and joined her brethren to counsel and act with them for crushed humanity, and I would not receive her aid, because it was "contrary to the usages of civilized society, and would throw a firebrand into the Anti-Slavery ranks."

H. C. WRIGHT

36

ANGELINA GRIMKÉ

Letter to Anne Warren Weston

Fort Lee, N.J., July 15, 1838

Angelina's letter to a leading member of the Boston Female Anti-Slavery Society inquired about the split within the movement and described the sisters' "domestic characters."

MY DEAR ANNA:
. . . I thank thee also for thy account of the N E Convention in reference to the Woman Question. Like all other truth, when brought out *practically,* it is causing deep searchings of heart & revealing the secrets of the soul. I believe this can no more be driven back from the field of investigation than the doctrine of Human rights, of which it is a part, & a very important part. And N E will be the battleground, for she is most certainly the moral light house of our nation. Perhaps it was all for the best Abby [Kelley] had to stand alone. I know how strengthening it is to feel that we have no arms of flesh to lean on, & for her sake I rejoice in her loneliness. . . . I cannot help hoping [Abby Kelley] will yet come out as a lecturer in the cause of the poor slave. Such practical advocacy of the rights of woman are worth every thing to *every* reform, at least, so I believe.

Has unity of spirit been restored to the Female Society in Boston? I long to hear from your *striving together* in the faith of Abolition as you once did. What plans are you pursuing this year? What are you doing with your friends?

We keep no help & therefore are filling up "the appropriate sphere of woman" to admiration, in the kitchen with baking pans & pots & steamers &c., & in the parlor & chambers with the broom & the duster. Indeed, I think our enemies wld rejoice, could they only look in upon us from day to day & see us toiling in domestic life, instead of lecturing to *promiscuous* audiences. Now I verily believe that we are *thus* doing *as much* for the cause of woman as we did by public speaking. For it is ab-

---

Boston Public Library, MS. A.9.2. v.10, 38.

solutely necessary that we should show that we are *not* ruined as domestic characters, but so far from it, *as soon* as duty calls us home, we can & do rejoice in the release from public service, & are as anxious to make good bread as we ever were to deliver a good lecture. Our ignorance and inexperience often leads to mistakes & failures in the cooking department, but to Theodore's contented and cheerful mind everything is good and home is delightful. We all like doing without a [hired] girl very much indeed for we find that it is very sweet to serve one another in love, each bearing a part of the burden, & so by a division of labor, rendering it comparatively easy to get along. . . .

I REMAIN THINE — ANGELINA G. WELD

37

LYDIA MARIA CHILD

Letter to Angelina Grimké

Boston, September 2, 1839

Child's published letter reviewed the causes of the split within the antislavery movement and asserted her own moderate but firm position on women's rights.

DEAR FRIEND:

You ask me what I am thinking about the anti-slavery controversy, and whether I am not disheartened by recent divisions. . . .

With regard to the Woman Question, as it is termed . . . the Massachusetts Society have simply refused to take action upon it when the minority have urged them to do it. In the beginning, we were brought together by strong sympathy for the slave, without stopping to inquire about each other's religious opinions, or appropriate spheres. Then, women were hailed by acclamation as helpers in the great work. They joined societies, they labored diligently, and they stood against a scoffing world bravely.

"On the Present State of the Anti-Slavery Cause," *The Liberator*, Sept. 6, 1839.

When the two Grimkés came among us, impediments in the way of their lecturing straight-way arose, particularly among the clergy. The old theological argument from St. Paul was urged, and the Grimkés replied in their own defense. A strong feeling of hostility to woman's speaking in public had always been latent in the clergy, and this incident aroused it all over the country. The sisters found obstacles so multiplied in their path, that they considered the establishment of woman's freedom of vital importance to the anti-slavery cause. "Little can be done for the slave," said they, "while this prejudice blocks up the way." They urged me to say and do more about woman's rights, nay, at times they gently rebuked me for my want of zeal.

I replied, "It is best not to talk about our rights, but simply go forward and do whatsoever we deem a duty. In toiling for the freedom of others, we shall find our own." On this ground I have ever stood; and so have my anti-slavery sisters. Instead of forcing this "foreign topic" into anti-slavery meetings or papers, we have sedulously avoided it. The Liberator has not meddled with the discussion, except when attacks upon the Grimkés seemed to render replies on their part absolutely necessary.

From that day to this, the clergy, as a body, have been extremely sensitive on the subject. Different minds assign different causes for this sensitiveness. Some respect it, as occasioned by a conscientious interpretation of Scripture; others consider it an honest but narrow bigotry; while others smiling say, "They are afraid the women will preach better and charge less." Without imputing motives, I simply state an obvious fact.

If there are clergymen, or others, in our ranks, who conscientiously believe it wrong for woman freely to utter her thoughts and impart her knowledge to any body who can derive benefit from the same, I should be the last to put any constraint on his opinions. . . .

From the beginning, women, by paying their money, have become members of Anti-Slavery societies and conventions in various free states. They have behaved with discretion and zeal, and been proverbially lavish of exertion. We claim no authority to prescribe or limit their mode of action, any more than we do that of other members. . . .

For my individual self, I now, as ever, would avoid any discussion of the woman question in Anti-Slavery meetings, or papers. But when a man advises me to withdraw from a society or convention, or not to act there according to the dictates of my own judgment, I am constrained to reply, "Thou canst not touch the freedom of my soul. I deem that I have duties to perform here. I make no onset upon your opinions and prejudices; but my moral responsibility lies between God and my own conscience. No human being can have jurisdiction over that."

But, my dear friend, these questions of Non-Resistants, Woman's Rights, &c. are only urged to effect a secret purpose far more important in the eyes of our opponents; viz. to get Garrison formally disowned by the abolitionists of Massachusetts. The causes which lead to this desire lie deep and spread wide. Many men can talk of the necessity of an Isaiah among the Jewish priests, who are unwilling to acknowledge the need of a prophet among a time-serving priesthood now.

To your second question, I answer that I am not discouraged by these dissensions. Disagreeable they unquestionably are; so much so, that we would be willing to give up anything but principle to avoid them; but, under God's Providence, they will mightily promote the cause of general freedom. . . .

God bless the Massachusetts Anti-Slavery Society! Good men and true women from the beginning until now!

YOURS AFFECTIONATELY,
L. MARIA CHILD

38

THE BOSTON FEMALE ANTI-SLAVERY SOCIETY

Annual Meeting

October 1839

This report offers a window onto the struggles within women's organizations in 1839, as women chose sides in the break between the AASS and the "new organization."

. . . The regular time for the annual meeting, occurred on the second Wednesday of October. A large number of women assembled, and after the usual devotional services, the Annual Report was read and accepted. There was in it no allusion relative to the present aspect of the antislavery cause. . . . By the Treasurer's Report it appeared that the proceeds of the society were less by about $300 than in 1837, a natural result

The Liberator, Nov. 1, 1839.

of the division introduced into the society by the Clerical Appeal, which division has been increasing up to the present time. . . . [T]he election of officers came on, Miss Parker presiding. Mrs. Child nominated Mrs. Southwick for President, and immediately some one nominated Miss Mary S. Parker. Mrs. Child said, that in the present state of the Society, we needed to have an impartial President. She would say, in all kindness and courtesy to Miss Parker, that this was not the case at present. She had been pained at the last meeting by the great want of impartiality with which the duties of the chair had been discharged.

Miss Anne Weston said, that though coinciding with Mrs. Child, in her view of the manner in which the President had presided at a former meeting, yet her reasons for being warmly opposed to Miss Parker's re-election were more important and weighty. They were reasons that applied to a majority of the last Board, namely, the fact that through its Board, this Society was placed in the attitude of sympathizing with and indirectly countenancing the new organization, and laboring to injure, as far as might be in its power, the Massachusetts A. S. Society. Believing the latter Society to be the best existing instrumentality that the anti-slavery cause possessed, she earnestly hoped that all who were true abolitionists, and desired to promote the best interests of the slave, would vote against Miss Parker's election.

Miss Lucy Ball appealed to the Society to sustain Miss Parker on the ground of her former services. With regard to the new organization, she asked, "Have we not the right to sympathize with it?"

Miss Parker said she knew partiality to be her chief failing, and she was obliged to Mrs. Child for pointing it out, though she must say Mrs. Child could hardly have been more personal. . . .

A number of members attempted to speak, and calls for the vote to be fairly taken were uttered from all parts of the room. Mrs. Southwick in a firm tone said—"I protest against this whole proceeding, and move that Miss Parker leave the chair, and Miss Sullivan take it, that the business of this meeting may be properly transacted." Mrs. Southwick was heard,—though all the time she was speaking, Miss Parker uttered incessant cries of there is no motion before the house. "Sit down, take your seat, Mrs. Southwick. You shall sit down." After Mrs. Southwick had so far succeeded in attracting Miss Parker's notice as to make her comprehend that she was submitting as a motion, that Miss Sullivan should be called to the chair, in the same hurried manner that characterized all she did during the latter portion of the afternoon, Miss Parker submitted the motion, declaring almost at the same instant, "It is not a vote." She then said—"Miss Mary Parker is elected President." . . .

The vote was taken by rising, and Miss Parker declared it to be carried. The vote was doubted, and though a number of ladies had left the house, and though the members seemed nearly equally divided, no counting was permitted. One lady said to Miss Parker, "I doubt the vote." "Take your seat," said she. "But I doubt the vote." "Then you may doubt it to the day of your death," was Miss Parker's rejoinder. . . . The pretended election of the other officers was carried on in an equally unconstitutional manner. . . . An adjournment was moved, and declared by Miss Parker to be carried.

What should be the course of faithful members of the Society at this time? The principles of the Society are inviolate. While they continue so, let us seek to save it from destruction. The ground on which we stand is holy. Let us not leave it, but contend for it, till falsehood and duplicity are fully exposed.

A LIFE-MEMBER OF THE BOSTON F. A. S. SOCIETY.
[ANNE WARREN WESTON]

AN INDEPENDENT WOMEN'S RIGHTS MOVEMENT IS BORN, 1840–1858

39

ELIZABETH CADY STANTON

On Meeting Lucretia Mott

London, June 1840

Differences between Stanton and Mott, which included Mott's being a generation older, seem to have drawn the two together. Stanton's account tells us about her initiation into the world of reform as well as about the impact Mott made on her.

. . . In June, 1840, I met Mrs. Mott for the first time, in London. Crossing the Atlantic in company with James G. Birney, then the Liberty Party

Elizabeth Cady Stanton, Susan B. Anthony, and Matilda Joslyn Gage, *History of Woman Suffrage, 1: 1848–1861* (New York: Fowler & Wells, 1881), 419–23.

candidate for President, soon after the bitter schism in the Anti-Slavery ranks, he described to me as we walked the deck, day after day, the women who had fanned the flames of dissension, and had completely demoralized the Anti-Slavery ranks. As my first view of Mrs. Mott was through his prejudices, no prepossessions in her favor biased my judgement. When first introduced to her at our hotel in Great Queen Street, with the other ladies from Boston and Philadelphia, who were delegates to the World's Convention, I felt somewhat embarrassed, as I was the only lady present who represented the "Birney faction," though I really knew nothing of the merits of the division, having been outside the world of reforms. Still, as my husband, and my cousin Gerrit Smith, were on that side, I supposed they would all have a feeling of hostility toward me. However, Mrs. Mott, in her sweet, gentle way, received me with great cordiality and courtesy, and I was seated by her side at dinner.

No sooner were the viands fairly dispensed, than several Baptist ministers began to rally the ladies on having set the abolitionists by the ears in America, and now proposing to do the same thing in England. I soon found that the pending battle was on woman's rights, and that, unwittingly, I was by marriage on the wrong side. As I had thought much on this question in regard to the laws, church action, and social usages, I found myself in full accord with the other ladies, combating most of the gentlemen at the table. . . . Calmly and skillfully Mrs. Mott parried all their attacks, now by her quiet humor turning the laugh on them, and then by her earnestness and dignity silencing their ridicule and sneers. I shall never forget the look of recognition she gave me when she saw, by my remarks, that I comprehended the problem of woman's rights and wrongs. How beautiful she looked to me that day. . . .

Mrs. Mott was to me an entirely new revelation of womanhood. I sought every opportunity to be at her side, and continually plied her with questions, and I shall never cease to be grateful for the patience and seeming pleasure, with which she fed my hungering soul. On one occasion, with a large party, we visited the British Museum, where it is supposed all people go to see the wonders of the world. On entering, Mrs. Mott and myself sat down near the door to rest for a few moments telling the party to go on, that we would follow. They accordingly explored all the departments of curiosities, supposing we were slowly following at a distance; but when they returned, there we sat in the same spot, having seen nothing but each other, wholly absorbed in questions of theology and social life. She had told me of the doctrines and divisions among "Friends"; of the inward light; of Mary Wollstonecraft, her social theories, and her demands of equality for women. I had been reading Combe's "Constitution of Man," and "Moral Philosophy," Channing's works, and Mary Wollstonecraft, though all tabooed by orthodox teachers; but I had never heard a woman talk what, as a Scotch Presbyterian, I had scarcely dared to think.

On the following Sunday I went to hear Mrs. Mott preach in a Unitarian church. Though I had never heard a woman speak, yet I had long believed she had the right to do so, and had often expressed the idea in private circles; but when at last I saw a woman rise up in the pulpit and preach earnestly and impressively, as Mrs. Mott always did, it seemed to me like the realization of an oft-repeated, happy dream. The day we visited the Zoological Gardens, as we were admiring the gorgeous plummage of some beautiful birds, one of our gentleman opponents remarked, "You see, Mrs. Mott, our Heavenly Father believes in bright colors. How much would it take from our pleasure, if all the birds were dressed in drab." "Yes," said she, "but immortal beings do not depend on their feathers for their attraction. With the infinite variety of the human face and form, of thought, feeling, and affection, we do not need gorgeous apparel to distinguish us. Moreover, if it is fitting that woman should dress in every color of the rainbow, why not man also? Clergymen, with their black clothes and white cravats, are quite as monotonous as Quakers." . . .

I remember on one occasion the entire American delegation were invited to dine with Samuel Gurney, a rich Quaker banker. He had an elegant place, a little out of London. The Duchess of Sutherland and Lord Morpeth, who had watched our anti-slavery struggle in this country with great interest, were quite desirous of meeting the American Abolitionists, and had expressed the wish to call on them at this time. Standing near Mrs. Mott when the coach and four gray horses with the six outriders drove up, Mr. Gurney, in great trepidation, said, "What shall I do with the Duchess?" "Give her your arm," said Mrs. Mott, "and introduce her to each member of the delegation." A suggestion no commoner in England would have presumed to follow. When the Duchess was presented to Mrs. Mott, her gracious ease was fully equaled by that of the simple Quaker woman. Oblivious to all distinctions of rank, she talked freely and wisely on many topics, and proved herself in manner and conversation the peer of the first woman in England. Mrs. Mott did not manifest the slightest restraint or embarrassment during that marked social occasion. No fictitious superiority ever oppressed her, neither did she descend in familiar surroundings from her natural dignity, but always maintained the perfect equilibrium of respect for herself and others.

I found in this new friend a woman emancipated from all faith in man-made creeds, from all fear of his denunciations. Nothing was too sacred for her to question, as to its rightfulness in principle and practice. "Truth for authority, not authority for truth," was not only the motto of her life, but it was the fixed mental habit in which she most rigidly held herself. It seemed to me like meeting a being from some larger planet, to find a woman who dared to question the opinions of Popes, Kings, Synods, Parliaments, with the same freedom that she would criticise an editorial in the *London Times,* recognizing no higher authority than the judgment of a pure-minded, educated woman. When I first heard from the lips of Lucretia Mott that I had the same right to think for myself that Luther, Calvin, and John Knox had, and the same right to be guided by my own convictions, and would no doubt live a higher, happier life than if guided by theirs, I felt at once a new-born sense of dignity and freedom; it was like suddenly coming into the rays of the noon-day sun, after wandering with a rushlight in the caves of the earth. When I confessed to her my great enjoyment in works of fiction, dramatic performances, and dancing, and feared from underneath that Quaker bonnet (I now loved so well) would come some platitudes on the demoralizing influence of such frivolities, she smiled, and said, "I regard dancing a very harmless amusement"; and added, "the evangelical Alliance that so readily passed a resolution declaring dancing a sin for a church member, tabled a resolution declaring slavery a sin for a bishop."

Sitting alone one day, as we were about to separate in London, I expressed to her my great satisfaction in her acquaintance, and thanked her for the many religious doubts and fears she had banished from my mind. She said, "There is a broad distinction between religion and theology. The one is a natural, human experience common to all well-organized minds. The other is a system of speculations about the unseen and the unknowable, which the human mind has no power to grasp or explain, and these speculations vary with every sect, age, and type of civilization. No one knows any more of what lies beyond our sphere of action than thou and I, and we know nothing." Everything she said seemed to me so true and rational, that I accepted her words of wisdom with the same confiding satisfaction that did the faithful Crito those of his beloved Socrates. And yet this pure, grand woman was shunned and feared by the Orthodox Friends throughout England. While in London a rich young Quaker of bigoted tendencies, who made several breakfast and tea parties for the American delegates, always omitted to invite Mrs. Mott. He very politely said to her on one occasion when he was inviting others in her presence, "Thou must excuse me, Lucretia, for not inviting thee with the rest, but I fear thy influence on my children!!"

40

ELIZABETH CADY STANTON

Letter to Sarah Grimké and Angelina Grimké Weld

London, June 25, 1840

Stanton's letter depicts her ability to side with the Garrisonian women even though her husband had joined the "new organization."

DEAR SISTERS, SARAH & ANGELINA:

Yesterday the convention closed, & I hasten to redeem my promise, to tell you something about it. We send you papers containing a minute account of all the proceedings, therefore I shall be very general in what I write. All things considered the convention has passed off more smoothly than any of us anticipated. The woman's rights question besides monopolizing one whole day has by being often referred to, created some little discord, for on this point we find a difference of opinion among the men & women here as well as with us in America. Garrison arrived on the fourth day of the meeting, but as the female delegates were not received and were not permitted to take their seats as delegates, he refused to take his, consequently his voice was not heard throughout the meeting. However last evening he opened his mouth, & forth came, in my opinion, much folly. . . . a feast was spread for us after which instead of toasts & jokes we had antislavery speeches. Garrison took this opportunity to relieve his full heart. This being a social occasion, & in no way connected with the convention he thought he might do so without sacrificing principle. This was his first appearance & he was received with many cheers, but oh! how soon by his want of judgement did he change the current of feeling in his audience,—a general look of disappointment was visible among the English ere he had spoken long. The chairman (for these social meetings which are frequent here are conducted with great order) was obliged to call him to order for wandering from the subject of conversation, which by general consent was to be slavery. Garrison touched, dwelt I might rather say, on woman's rights, poor laws, temperance, &c &c. . . .

Weld-Grimké Papers.

Lucretia Mott has just given me a long message for you, which condensed is that she thinks you have both been in a state of retiracy long enough, & that it is not right for you to be still, longer, that you should either write for the public or speak out for *oppressed woman.* Sarah in particular she thinks should appear in public again as she has no duties to prevent her. She says a great struggle is at hand & that all the friends of freedom for woman must rally round the *Garrison standard.* I have had much conversation with Lucretia Mott & I think her a peerless woman. She has a clear head & warm heart—her views are many of them so new & strange that my *causality* finds great delight in her society. The quakers here have not all received her cordially, they fear her heretical notions. I am often asked if you have not changed your opinions on woman's rights & I have invariably taken the liberty to say no. . . .

Your names are always mentioned with great enthusiasm. You would laugh I am sure to see the look of surprize when to the list of virtues I add your great skill in discharging all your domestic avocations. Dear friends how much I love you!! What a trio! for me to love. You have no idea what a hold you have on my heart. The two green spots to me in America are the peaceful abodes of cousin Gerrit [Smith] & Theodore Weld. oh! I cannot tell you with what delight I look forward to the many hours I hope to spend in those places of pleasant memories. . . .

Henry wishes me to say that he attributes his freedom from seasickness to *his strict observance of the Graham system.* . . .

YOURS IN LOVE
ELIZABETH C. STANTON

41

ELIZABETH CADY STANTON

Planning the Seneca Falls Convention

1848

A casual social visit turned into a serious planning meeting for the first women's rights convention in the western world. Elizabeth Cady Stanton's

Stanton, et al., *History of Woman Suffrage,* 1:67–69.

account of that meeting reveals the improvisation that produced the convention's chief document, the "Declaration of Sentiments."

WOMAN'S RIGHTS CONVENTION.—A Convention to discuss the social, civil, and religious condition and rights of woman, will be held in the Wesleyan Chapel, at Seneca Falls, N.Y., on Wednesday and Thursday, the 19th and 20th of July, current; commencing at 10 o'clock A.M. During the first day the meeting will be exclusively for women, who are earnestly invited to attend. The public generally are invited to be present on the second day, when Lucretia Mott, of Philadelphia, and her other ladies and gentlemen, will address the convention.

This call, without signature, was issued by Lucretia Mott, Martha C. Wright, Elizabeth Cady Stanton, and Mary Ann McClintock. At this time Mrs. Mott was visiting her sister Mrs. Wright, at Auburn, and attending the Yearly Meeting of Friends in Western New York. Mrs. Stanton, having recently removed from Boston to Seneca Falls, finding the most congenial association of Quaker families, met Mrs. Mott incidentally for the first time since her residence there. They at once returned to the topic they had so often discussed, walking arm in arm in the streets of London, and Boston, "the propriety of holding a woman's convention." These four ladies, sitting round the tea-table of Richard Hunt, a prominent Friend near Waterloo, decided to put their long-talked-of resolution into action, and before the twilight deepened into night, the call was written, and sent to the *Seneca County Courier.* On Sunday morning they met in Mrs. McClintock's parlor to write their declaration, resolutions, and to consider subjects for speeches. As the convention was to assemble in three days, the time was short for such productions; but having no experience in the *modus operandi* of getting up conventions, nor in that kind of literature, they were quite innocent of the herculean labors they proposed. On the first attempt to frame a resolution; to crowd a complete thought, clearly and concisely, into three lines; they felt as helpless and hopeless as if they had been suddenly asked to construct a steam engine. And the humiliating fact may as well now be recorded that before taking the initiative step, those ladies resigned themselves to a faithful perusal of masculine productions. The reports of Peace, Temperance, and Anti-Slavery conventions were examined, but all alike seemed too tame and pacific for the inauguration of a rebellion such as the world had never before seen. They knew women had wrongs, but how to state them was the difficulty, and this was increased from the fact that they themselves were fortunately organized and conditioned; they were neither "sour old maids," childless women," nor "divorced wives," as the newspapers declared them to be. While they had felt the insults incident to sex, in many ways, as every proud, thinking woman must, in the laws, religion, and literature of the world, and in the invidious and degrading sentiments and customs of all nations, yet they had not in their own experience endured the coarser forms of tyranny resulting from unjust laws, or association with immoral and unscrupulous men, but they had souls large enough to feel the wrongs of others, without being sacrificed in their own flesh.

After much delay, one of the circle took up the Declaration of 1776, and read it aloud with much spirit and emphasis, and it was at once decided to adopt the historic document, with some slight changes such as substituting "all men" for "King George." Knowing that women must have more to complain of than men under any circumstances possibly could, and seeing the Fathers had eighteen grievances, a protracted search was made through statute books, church usages, and customs of society to find that exact number. Several well-disposed men assisted in collecting the grievances, until, with the announcement of the eighteenth, the women felt they had enough to go before the world with a good case. One youthful lord remarked, "Your grievances must be grievous indeed, when you are obliged to go to books in order to find them out."

42

Report of the Woman's Rights Convention

Seneca Falls, N.Y., July 19–20, 1848

The length and complexity of this report reflected the strength of the reform culture that created it. Those who attended the Seneca Falls Convention were experienced conference goers. They knew how to create a community

Report of the Woman's Rights Convention, Held at Seneca Falls, N.Y., July 19th and 20th, 1848 (Rochester, 1848); italicized portion from Stanton et al., ***History of Woman Suffrage,*** 1:73.

of concerns from diverse individual perspectives, and how to gain a hearing for those concerns within the broader society.

A Convention to discuss the Social, Civil, and Religious Condition of Woman, was called by the Women of Seneca County, N.Y., and held at the village of Seneca Falls, in the Wesleyan Chapel, on the 19th and 20th of July, 1848.

The question was discussed throughout two entire days: the first day by women exclusively, the second day men participated in the deliberations. Lucretia Mott, of Philadelphia, was the moving spirit of the occasion.

On the morning of the 19th, the Convention assembled at 11 o'clock. The meeting was organized by appointing Mary M'Clintock Secretary. The object of the meeting was then stated by Elizabeth C. Stanton; after which, remarks were made by Lucretia Mott, urging the women present to throw aside the trammels of education, and not allow their new position to prevent them from joining in the debates of the meeting. The Declaration of Sentiments, offered for the acceptance of the Convention, was then read by E. C. Stanton. A proposition was made to have it re-read by paragraph, and after much consideration, some changes were suggested and adopted. The propriety of obtaining the signatures of men to the Declaration was discussed in an animated manner: a vote in favor was given; but concluding that the final decision would be the legitimate business of the next day, it was referred.

Adjourned to half-past two.

In the afternoon, the meeting assembled according to adjournment, and was opened by reading the minutes of the morning session. E. C. Stanton then addressed the meeting, and was followed by Lucretia Mott. The reading of the Declaration was called for, an addition having been inserted since the morning session. A vote taken upon the amendment was carried, and papers circulated to obtain signatures. The following resolutions were then read:

The following resolutions were discussed by Lucretia Mott, Thomas and Mary Ann McClintock, Amy Post, Catherine A. F. Stebbins, and others, and were adopted.

WHEREAS, The great precept of nature is conceded to be, that "man shall pursue his own true and substantial happiness." Blackstone in his Commentaries remarks, that this law of Nature being coeval with mankind, and dictated by God himself, is of course superior in obligation to any other. It is binding over all the globe, in all countries and at all times;

no human laws are of any validity if contrary to this, and such of them as are valid, derive all their force, and all their validity, and all their authority, mediately and immediately, from this original;
Therefore,

Resolved, That such laws as conflict, in any way, with the true and substantial happiness of woman, are contrary to the great precept of nature and of no validity, for this is "superior in obligation to any other."

Resolved, That all laws which prevent woman from occupying such a station in society as her conscience shall dictate, or which place her in a position inferior to that of man, are contrary to the great precept of nature, and therefore of no force or authority.

Resolved, That woman is man's equal—was intended to be so by the Creator, and the highest good of the race demands that she should be recognized as such.

Resolved, That the women of this country ought to be enlightened in regard to the laws under which they live, that they may no longer publish their degradation, by declaring themselves satisfied with their present position, nor their ignorance, by asserting that they have all the rights they want.

Resolved, That inasmuch as man, while claiming for himself intellectual superiority, does accord to woman moral superiority, it is preeminently his duty to encourage her to speak and teach, as she has an opportunity, in all religious assemblies.

Resolved, That the same amount of virtue, delicacy, and refinement of behavior, that is required of woman in the social state, should also be required of man, and the same transgressions should be visited with equal severity on both man and woman.

Resolved, That the objection of indelicacy and impropriety, which is so often brought against woman when she addresses a public audience, comes with a very ill-grace from those who encourage, by their attendance, her appearance on the stage, in the concert, or in feats of the circus.

Resolved, That woman has too long rested satisfied in the circumscribed limits which corrupt customs and a perverted application of the Scriptures have marked out for her, and that it is time she should move in the enlarged sphere which her great Creator has assigned her.

Resolved, That it is the duty of women of this country to secure themselves their sacred right to the elective franchise.

Resolved, That the equality of human rights results necessarily from the fact of the identity of the race in capabilities and responsibilities.

Resolved, therefore, That, being invested by the Creator with the same capabilities, and the same consciousness of responsibility for their exercise, it is demonstrably the right and duty of woman, equally with man, to promote every righteous cause by every righteous means; and especially in regard to the great subjects of morals and religion, it is self-evidently her right to participate with her brother in teaching them, both in the private and public, by writing and by speaking, by any instrumentalities proper to be used, and in any assemblies proper to be held; and this being a self-evident truth growing out of the divinely implanted principles of human nature; any custom or authority adverse to it, whether modern or wearing the hoary sanction of antiquity, is to be regarded as a self-evident falsehood, and at war with the interests of mankind.

Lucretia Mott read a humorous article from a newspaper, written by Martha C. Wright. After an address by E. W. M'Clintock, the meeting adjourned to 10 o'clock the next morning.

In the evening, Lucretia Mott spoke with her usual eloquence and power to a large and intelligent audience on the subject of reforms in general.

Thursday Morning

The Convention assembled at the hour appoint, James Mott, of Philadelphia, in the Chair. The minutes of the previous day having been read, E. C. Stanton again read the Declaration of Sentiments, which was freely discussed by Lucretia Mott, Ansel Bascom, S. E. Woodworth, Thomas and Mary Ann M'Clintock, Frederick Douglass, Amy Post, Catharine Stebbins, and Elizabeth C. Stanton, and was unanimously adopted, as follows:

Declaration of Sentiments

When, in the course of human events, it becomes necessary for one portion of the family of man to assume among the people of the earth a position different from that which they have hitherto occupied, but one to which the laws of nature and of nature's God entitle them, a decent respect to the opinions of mankind requires that they should declare the causes that impel them to such a course.

We hold these truths to be self-evident: that all men and women are created equal: that they are endowed by their Creator with certain inalienable rights; that among these are life, liberty, and the pursuit of happiness; that to secure these rights governments are instituted, deriving their just powers from the consent of the governed. Whenever any form of government becomes destructive of these ends, it is the right of those who suffer from it to refuse allegiance to it, and to insist upon the institution of a new government, laying its foundation on such principles and organizing its powers in such form, as to them shall seem most likely to effect their safety and happiness. Prudence, indeed, will dictate that governments long established should not be changed for light and transient causes; and accordingly all experience hath shown that mankind are more disposed to suffer, while evils are sufferable, than to right themselves by abolishing the forms to which they were accustomed. But when a long train of abuses and usurpations, pursuing invariably the same object evinces a design to reduce them under absolute despotism, it is their duty to throw off such government, and to provide new guards for their future security. Such has been the patient sufferance of the women under this government, and such is now the necessity which constrains them to demand the equal station to which they are now entitled.

The history of mankind is a history of repeated injuries and usurpations on the part of man toward woman, having in direct object the establishment of an absolute tyranny over her. To prove this, let facts be submitted to a candid world.

He has never permitted her to exercise her inalienable right to the elective franchise.

He has compelled her to submit to laws, in the formation of which she had no voice.

He has withheld from her rights which are given to the most ignorant and degraded men — both natives and foreigners.

Having deprived her of this first right of a citizen, the elective franchise, thereby leaving her without representation in the halls of legislation, he has oppressed her on all sides.

He has made her, if married, in the eye of the law, civilly dead.

He has taken from her all right to property, even to the wages she earns.

He has made her, morally, an irresponsible being, as she can commit many crimes with impunity, provided they be done in the presence of her husband. In the covenant of marriage, she is compelled to promise obedience to her husband, he becoming, to all intents and purposes, her master — the law giving him the power to deprive her of her liberty, and to administer chastisement.

He has so framed the laws of divorce, as to what shall be the proper causes, and in case of separation, to whom the guardianship of the children shall be given, as to be wholly regardless of the happiness of women — the law, in all cases, going upon a false supposition of the supremacy of man, and giving all power into his hands.

After depriving her of all rights as a married woman, if single, and the owner of property, he has taxed her to support a government which recognizes her only when her property can be made profitable to it.

He has monopolized nearly all the profitable employments, and from those she is permitted to follow, she receives but a scanty remuneration.

He closes against her all the avenues to wealth and distinction which he considers most honorable to himself. As a teacher of theology, medicine, or law, she is not known.

He has denied her the facilities for obtaining a thorough education — all colleges being closed against her.

He allows her in Church, as well as State, but in a subordinate position, claiming Apostolic authority for her exclusion from the ministry, and with some exceptions, from a public participation in the affairs of the Church.

He has created a false public sentiment by giving to the world a different code of morals for men and women, by which moral delinquencies which exclude women from society, are not only tolerated, but deemed of little account in man.

He has usurped the prerogative of Jehovah himself, claiming it as his right to assign her a sphere of action, when that belongs to her conscience and to her God.

He has endeavored, in every way that he could, to destroy her confidence in her own powers, to lessen her self-respect, and to make her willing to lead a dependent and abject life.

Now in view of this entire disfranchisement of one-half the people of this country, their social and religious degradation — in view of the unjust laws above mentioned, and because women do feel themselves aggrieved, oppressed, and fraudulently deprived of their most sacred rights, we insist that they have immediate admission to all the rights and privileges which belong to them as citizens of the United States.

In entering upon the great work before us, we anticipate no small amount of misconception, misrepresentation, and ridicule; but we shall use every instrumentality within our power to effect our object. We shall employ agents, circulate tracts, petition the State and National legislatures, and endeavor to enlist the pulpit and the press in our behalf. We hope this Convention will be followed by a series of Conventions embracing every part of the country.

Firmly relying upon the final triumph of the Right and the True, we do this day affix our signatures to this declaration. [signatures of 68 women, Lucretia Mott's first]

The following are the names of the gentlemen present in favor of the movement: [signatures of 32 men, including Frederick Douglass.]

The meeting adjourned until two o'clock.

Afternoon Session

At the appointed hour the meeting convened. The minutes having been read, the resolutions of the day before were read and taken up separately. Some, from their self-evident truth, elicited but little remark; others, after some criticism, much debate, and some slight alterations, were finally passed by a large majority. . . .

[*The only resolution that was not unanimously adopted was the ninth, urging women of the country to secure themselves the elective franchise. Those who took part in the debate feared a demand for the right to vote would defeat others they deemed more rational, and make the whole movement ridiculous.*

But Mrs. Stanton and Frederick Douglass, seeing that the power to choose rulers and make laws was the right by which all others could be secured, persistently advocated the resolution, and at last carried it by a small majority.]

The meeting closed with a forcible speech from Lucretia Mott.

Adjourned to half-past seven o'clock.

Evening Session

The meeting opened by reading the minutes, Thomas M'Clintock in the Chair. As there had been no opposition expressed during the Convention to this movement, and although, after repeated invitations, no objections had presented themselves, E. C. Stanton volunteered an address in defence of the many severe accusations brought against the much-abused "Lords of Creation."

Thomas M'Clintock then read several extracts from Blackstone, in proof of woman's servitude to man; after which Lucretia Mott offered and spoke to the following resolution:

Resolved, That the speedy success of our cause depends upon the zealous and untiring efforts of both men and women, for the overthrow of the monopoly of the pulpit, and for the securing to woman an equal participation with men in the various trades, professions, and commerce. The Resolution was adopted.

M. A. M'Clintock, Jr. delivered a short, but impressive address, calling upon woman to arouse from her lethargy and be true to herself and her God. When she had concluded, Frederick Douglass arose, and in an excellent and appropriate speech, ably supported the cause of woman.

The meeting was closed by one of Lucretia Mott's most beautiful and spiritual appeals. She commanded the earnest attention of that large audience for nearly an hour.

43

SOJOURNER TRUTH

Speech at Akron Women's Rights Convention

Ohio, June 1851

Reprinted in a variety of forms, this notable speech by Sojourner Truth — seer, abolitionist, and woman's rights advocate — was most accurately rendered in the Anti-Slavery Bugle *of Salem, Ohio. There it appeared with an introductory paragraph.*

One of the most unique and interesting speeches of the Convention was made by Sojourner Truth, an emancipated slave. It is impossible to transfer it to paper, or convey any adequate idea of the effect it produced upon the audience. Those only can appreciate it who saw her powerful form, her whole-souled, earnest gesture, and listened to her strong and truthful tones. She came forward to the platform and addressing the President said with great simplicity:

May I say a few words? Receiving an affirmative answer, she proceeded; I want to say a few words about this matter. I am a woman's rights [*sic*]. I have as much muscle as any man, and can do as much work as any man. I have plowed and reaped and husked and chopped and mowed, and can any man do more than that? I have heard much about the sexes being equal; I can carry as much as any man, and eat as much too, if I can get it. I am as strong as any man that is now.

Anti-Slavery Bugle, Salem, Ohio, June 21, 1851, reprinted in Carleton Mabee, *Sojourner Truth: Slave, Prophet, Legend* (New York: New York University Press, 1993), 81.

As for intellect, all I can say is, if woman have a pint and man a quart — why can't she have her little pint full? You need not be afraid to give us our rights for fear we will take too much — for we won't take more than our pint'll hold.

The poor men seem to be all in confusion and don't know what to do. Why children, if you have woman's rights give it to her and you will feel better. You will have your own rights, and they won't be so much trouble.

I can't read, but I can hear. I have heard the Bible and have learned that Eve caused man to sin. Well if woman upset the world, do give her a chance to set it right side up again. The lady has spoked about Jesus, how he never spurned woman from him, and she was right. When Lazarus died, Mary and Martha came to him with faith and love and besought him to raise their brother. And Jesus wept — and Lazarus came forth. And how came Jesus into the world? Through God who created him and woman who bore him. Man, where is your part?

But the women are coming up bless be God and a few of the men are coming up with them. But man is in a tight place, the poor slave is on him, woman is coming on him, and he is surely between a hawk and a buzzard.

44

ABBY H. PRICE

Address to the "Woman's Rights Convention"

Worcester, Mass., October 1850

Abby Price exemplified the increasingly secular language used to advocate women's rights in the 1850s.

In our account of the work of Creation, when it was so gloriously finished in the garden of Eden, by placing there, in equal companionship, man and woman, made in the image of God, alike gifted with intellect, alike

Proceedings of the Woman's Rights Convention Held at Worcester, October 23rd and 24th, 1850 (Boston: Prentiss & Sawyer, 1851).

endowed with immortality, it is said the Creator looked upon his work, and pronounced it good — that "the morning stars sang together, and all the sons of God shouted for joy." Since that time, through the slow rolling of darkened ages, man has ruled by physical power, and wherever he could gain the ascendancy, there he has felt the right to dictate — even though it degraded his equal companion — the mother who bore him — the playmate of his childhood — the daughter of his love. Thus, in many countries we see woman reduced to the condition of a slave, and compelled to do all the drudgery necessary to her lord's subsistence. In others she is dressed up as a mere plaything, for his amusement; but everywhere he has assumed to be her head and lawgiver, and only where Christianity has dawned, and right not might been the rule, has woman had anything like her true position. In this country even, republican, so called, and Christian, her rights are but imperfectly recognised, and she suffers under the disability of caste. These are the facts that in the light of the nineteenth century, demand our attention. "Are we always to remain in this position" is a question we have come here to discuss.

The natural rights of woman are co-equal with those of man. So God created man in his own image; in the image of God created he him; male and female, created he them. There is not one particle of difference intimated as existing between them. They were both made in the image of God. Dominion was given to both over every other creature, but not over each other. They were expected to exercise the viceregency given to them by their Maker in harmony and love.

In contending for this co-equality of woman's with man's rights, it is not necessary to argue, either that the sexes are by nature equally and indiscriminately adapted to the same positions and duties, or that they are absolutely equal in physical and intellectual ability; but only that they are absolutely equal in their rights to life, liberty, and the pursuit of happiness — in their rights to do, and to be, individually and socially, all they are capable of, and to attain the highest usefulness and happiness, obediently to the divine moral law.

These are every man's rights, of whatever race or nation, ability or situation, in life. These are equally every woman's rights, whatever her comparative capabilities may be — whatever her relations may be. These are human rights, equally inherent in male and female. To repress them in any degree is in the same degree usurpation, tyranny, and oppression. We hold these to be self-evident truths, and shall not now discuss them. We shall assume that happiness is the chief end of all human beings; that existence is valuable in proportion as happiness is promoted and secured; and that, on the whole, each of the sexes is equally necessary to the common happiness, and in one way or another is equally capable, with fair opportunity, of contributing to it. Therefore each has an equal right to pursue and enjoy it. This settled, we contend:

1. That women ought to have equal opportunities with men for suitable and well compensated employment.
2. That women ought to have equal opportunities, privileges, and securities with men for rendering themselves pecuniarily independent.
3. That women ought to have equal legal and political rights, franchises, and advantages with men. . . .

Human beings cannot attain true dignity or happiness except by true usefulness. This is true of women as of men. It is their duty, privilege, honor, and bliss to be useful. Therefore give them the opportunity and encouragement. If there are positions, duties, occupations, really unsuitable to females, as such, let these be left to males. If there are others unsuitable to men, let these be left to women. Let all the rest be equally open to both sexes. And let the compensation be graduated justly, to the real worth of the services rendered, irrespective of sex. . . .

What good reason is there why women should not be educated to mercantile pursuits, to engage in commerce, to invent, to construct, in fine [in sum] to do anything she can do? Why so separate the avocations of the sexes? I believe it impossible for woman to fulfil the design of God in her creation until her brethren mingle with her more as an equal, as a moral being, and lose in the dignity of her immortal nature the idea of her being a female. Until social intercourse is purified by the forgetfulness of sex we can never derive high benefit from each other's society in the active business of life. Man inflicts injury upon woman, unspeakable injury in placing her intellectual and moral nature in the background, and woman injures herself by submitting to be regarded only as a female. She is called upon loudly, by the progressive spirit of the age, to rise from the station where man, not God, has placed her, and to claim her rights as a moral and responsible being, equal with man.

As such, both have the same sphere of action, and the same duties devolve on both, though these may vary according to circumstances. Fathers and mothers have sacred duties and obligations devolving upon them which cannot belong to others. These do not attach to them as man and woman, but as parents, husbands, and wives. In all the majesty of moral power, in all the dignity of immortality let woman plant herself side by side with man on the broad platform of equal human rights. . . .

Our sisters, whose poverty is caused by the oppressions of society, who are driven to sin by want of bread,—then regarded with scorn and turned away from with contempt! I appeal to you in their behalf, my friends. Is it not time to throw open to women, equal resources with men, for obtaining honest employment? If the extremity of human wretchedness—a condition which combines within itself every element of suffering, mental and physical, circumstantial and intrinsic—is a passport to our compassion, every heart should bleed for the position of these poor sufferers. . . .

Let us arise then in all the majesty of renewed womanhood and say, we must be free. We will attend to our previous home duties faithfully, cheerfully, but we must do it voluntarily, in obedience to our Maker, who placed these responsibilities more especially upon us. If the affairs of the nation demand the attention of our fathers, our husbands, and our brothers, allow us to act with them for the right, according to the dictates of our own consciences. Then we will educate our sons and our daughters as equal companions, alike interested in whatever concerns the welfare of the race. Our daughters, equally provided for the serious business of life, shall no longer be dependent upon the chances of marriage; teaching them not to live wholly in their affections, we will provide for them, as for our sons, a refuge from the storms of life, by opening to them the regions of high intellectual culture, of pecuniary independence, and of moral and political responsibilities.

45

Proceedings of the Colored Convention

Cleveland, September 6, 1848

The first women's rights speaker in the Negro Convention Movement, Mrs. Sanford, probably a friend of Frederick Douglass from Rochester, said that women ask for rights "granted by a higher disposer of human events than man." These rights included married women's property rights, and (an indirect reference to woman suffrage) "to co-operate in making the

The North Star, 1, no. 40 (Sept. 29, 1848): 1.

laws we obey." After her speech, Frederick Douglass outmaneuvered the opponents of women's rights and the convention passed a resolution in favor of women's participation in future meetings.

. . . After an animated discussion upon the indefinite postponement, the Rules were suspended to hear remarks from a lady who wished to say something on the rights of Woman. The President then introduced to the audience, Mrs. Sanford, who made some eloquent remarks of which the following is a specimen:

> From the birthday of Eve, the then prototype of woman's destiny, to the flash of the star of Bethlehem, she had been the slave of power and passion. If raised by courage and ambition to the proud trial of heroism, she was still the marred model of her first innocence; if thrown by beauty into the ordeal of temptation, man lost his own dignity in contemning her intellectual weight, and refusing the right to exercise her moral powers; if led by inclination to the penitential life of a recluse, the celestial effulgence of a virtuous innocence was lost, and she lived out woman's degradation!
>
> But the day of her regeneration dawned. The Son of God had chosen a mother from among the daughters of Eve! A Saviour, who could have come into this a God-man ready to act, to suffer, and be crucified, came in the helplessness of infancy, for woman to cherish and direct. Her *exaltation was consummated!*
>
> True, we ask for the Elective Franchise; for right of property in the marriage covenant, whether earned or bequeathed. True, we pray to co-operate in making the laws we obey; but it is not to domineer, to dictate or assume. We ask it, for it is a right, granted by a higher disposer of human events than man. We pray for it now, for there are duties around us, and we weep at our inability.
>
> And to the delegates, officers, people and spirit of this Convention, I would say, God speed you in your efforts for elevation and freedom; stop not; shrink not, look not back, till you have justly secured an *unqualified citizenship of the United States, and those inalienable rights granted you by an impartial Creator.*

Convention passed a vote of thanks to Mrs. Sanford, and also requested a synopsis of her, from which the above are extracts.

A vote of thanks was here passed to John M. Sterling, Esq., of Cleveland, for the presentation of a bundle of books entitled "Slavery as it is."[1]

[1] Published by the AASS and edited primarily by Theodore Weld, *American Slavery as It Is* first appeared in 1839.

Discussion was resumed on the indefinite postponement of the Resolution as to Woman's Right. Objection was made to the resolution, and in favor of its postponement, by Messrs. Langston and Day, on the ground that we had passed one similar, making all colored persons present, delegates to this Convention, and they considered *women persons.*

Frederick Douglass moved to amend the 33d Resolution, by saying that the word persons used in the resolution designating delegates be understood to include *woman.* On the call for the previous question, the Resolution was not indefinitely postponed. Mr. Douglass's amendment was seconded and carried, with three cheers for woman's rights. . . .

[Resolutions]

33. Whereas, we fully believe in the equality of the sexes, therefore,

Resolved, That we hereby invite females hereafter to take part in our deliberations. . . .

46

"Woman's Rights"

October 1, 1849

This article exemplifies the many popular periodicals that promoted ideas about women's rights in the 1840s. A flourishing popular press provided many outlets for links between the emerging women's rights movement and other social movements. A women's temperance periodical, The Lily, ***was founded in January 1849 by Amelia Bloomer, a Seneca Falls writer who had attended the 1848 convention. In 1853 the magazine moved to Ohio and claimed a national readership of 6,000, but it ceased publication in 1856.***

Start not dear reader, as your eye rests upon the above words, nor think that we are going to nominate either you or ourself for the Governorship or the Presidency. No, it is not time to make lady Presidents yet, and for ourself we can say that we have no aspirations of the kind at present;—

The Lily, 1, 10 (Oct. 1, 1849).

but according to the belief of some, the day will soon come when woman may claim her "rights," in this respect, and then we may not be backward in taking a seat in the Presidential chair, provided the good people shall so will it.

It is not our right to hold office or to rule our country, that we would not advocate. Much, very much, must be done to elevate and improve the character and minds of our sex, before we are capable of ruling our own households as we ought, to say nothing of holding in our hands the reins of government. But woman has rights which she knows not of, or knowing, disregards. She has rights of which she is deprived — or rather, of which she deprives herself. She is willing to sit down within the narrow sphere assigned her by man, and make no effort to obtain her just rights, or free herself from the oppressions which are crushing her to the earth. She tamely submits to be governed by such laws as man sees fit to make and in making which she has no voice. We know that many of us think we have rights enough, and we are content with what we have; but we forget how many thousand wives and mothers worthy as ourselves, are compelled by the unjust laws of our land, to drag out a weary life and submit to indignities which no man would bear. It is stated that thirty thousand die annually from the effects of intoxicating drinks; an equal number of drunkards must stand ready to fall. Think of the wives and mothers of this great number — of their untold griefs — of their hidden sorrows — of their broken hearts — of their hunger and nakedness — their unwearied toil to procure a bare pittance to save their little ones from starvation — of the wretched life they lead, and the unmourned death they die. Think of all this, and then tell us not that woman has her rights. Many of the number thus destroyed inch by inch, have been reared amid all the luxuries that wealth and power can bestow. Many of them possess accomplishments that might have graced the most refined society, and who, had their lots been differently cast, would have been courted and sought after by those who now spurn them — and for what? Simply because they have been so unfortunate as to wed a drunkard — or rather because they upon whom they bestowed their young affections, and who vowed to love and protect them, have proved false to all their vows, and left them to the rough blasts of an unpitying world. What rights have the drunkard's wife and children? — Who listens to their tale of woe, or lends a pitying ear to their cry?

A woman is entitled to the same rights as a man, but does she have them? Dare men pretend that she does? What right have they to make laws which deprive her of every comfort, strip her of every friend, and doom her to a wretched existence? And yet they do this, and then if she dare to complain, and ask to be relieved from these tyranical laws, she is thought to be out of her place, and overstepping the bounds of female delicacy! This is why they so tamely submit to martyrdom by the laws. The statute book of this free country bears upon its leaves a foul stain called a license law [a license to sell alcoholic beverages]. By this law men are bidden to go forth and pursue a business which deprives thirty thousand annually of life — worse than murders twenty thousand wives and mothers, and sixty thousand children. For the privilege thus allowed, the law claims in return a few dollars from those who pursue this *moral* and *honorable* business! It is useless for our sex to seek redress at the hands of the law, from the cruel wrongs inflicted upon them, for it will give them none — it does not recognise their right to protection. But should they dare to raise their hand against their destroyers, and return injury for injury, then the law quickly defends its agents and metes out punishment for her who ventures to defend herself.

We ask not for the honors or emoluments of office for our sex, but we claim that they are unjustly deprived of their rights. We do not believe that man has the *right,* if he has the power, to make laws which will deprive us of any of the comforts of life — or if he does make the laws without our consent, we are not bound to obey them. Unless those who claim the power of legislating for us, will do something to ameliorate the condition of the down trodden victims of their cruel enactments, it is not only the right, but the duty of those trampled upon, to assert their claim to protection.

47

"Just Treatment of Licentious Men"

January 1838

Although nowhere so avidly defended or fully articulated as in the anti-slavery movement, women's rights ideas also appeared in other social movements, including moral reform. In this letter to the Friend of Virtue,

Friend of Virtue, Jan. 1838, 2–4; reprinted by Daniel Wright, "What Was the Appeal of Moral Reform to Antebellum Northern Women?" in "Women and Social Movements in the United States, 1830–1930," website at http://womhist.binghamton.edu.

the chief publication of the New England branch of the movement, the writer shows how women were taking the initiative in reforming sexual relations by condemning predatory male sexuality. Although these women tended to be much more conservative than Garrisonian abolitionist women in their views of women's public rights, they could be quite radical in their defense of women's right to control their own bodies and in their call for a single standard of sexual behavior for men and women.

DEAR SISTERS:
As members with us of the body of the Lord Jesus Christ, we take the liberty of addressing you on a subject near our hearts, and of the deepest interest to our sex. We ask your serious attention, while we press upon your consciences the inquiry, "Is it right to admit to the society of virtuous females, those unprincipled and licentious men, whose conduct is fraught with so much evil to those who stand in the relation to us of sisters?" True, God designed that man should be our protector, the guardian of our peace, our happiness, and our honor; but how often has he proved himself a traitor to his trust, and the worst enemy of our sex? The deepest degradation to which many of our sex have been reduced, the deepest injuries they have suffered, have been in consequence of his perfidy. He has betrayed, and robbed, and forsaken his victim, and left her to endure alone the untold horrors of a life embittered by self-reproach, conscious ignominy, and exclusion from every virtuous circle.

Is there a woman among us, whose heart has not been pained at the fall and fate of some one sister of her sex? Do you say the guilty deserve to suffer and must expect it? Granted. But why not let a part of this suffering fall on the destroyer? Why is he caressed and shielded from scorn by the countenance of the virtuous, and encouraged to commit other acts of perfidy and sin, while his victim, for one offence, is trampled upon, despised and banished from all virtuous society; The victim thus crushed, yields herself to despair, and becomes a practical illustration of the proverb that, "A bad woman is the worst of all God's creatures." Surely, if she is worse, after her fall, than man equally fallen, is there not reason to infer that in her nature there is something more chaste, more pure and refined, and exalted than in his? Is it then not worth while to do something to prevent her from becoming a prey to the perfidy and baseness of unprincipled man, and a disgrace to her sex? Do you ask, what can woman do, and reply as have some others, "We must leave this work for the men?" Can we expect the wolf, ravenous for his prey, to throw up a barrier to protect the defenceless sheep? As well might we expect this, as to expect that men as a body will take measures to redress the wrongs of woman.

Dear sisters, women have commenced this work, and women must see it carried through. Commenced by women? No it was commenced by one who is now, we trust, a sainted spirit in heaven, and who sacrificed his life in the cause. Yes, he fell a martyr in the conflict, but not till he had effectually roused the women of the nation to enlist in the cause he had commenced.[1] Moral Reform is the first of causes to our sex. It involves principles, which if faithfully and perseveringly applied, will preserve the rights and elevate the standing of our sex in society. As times have been, the libertine has found as ready a passport to the society of the virtuous, as any one, and he has as easily obtained a good wife, as the more virtuous man. But a new era has commenced. Woman has erected a standard, and laid down the principle, that man shall not trample her rights, and on the honor of her sex with impunity. She has undertaken to banish licentious men from all virtuous society. And mothers, wives, sisters, and daughters will you lend your influence to this cause? Prompt action in the form of association will accomplish this work. Females in this manner must combine their strength and exert their influence. Will you not join one of these bands of the pious? The cause has need of your interest, your prayers, and your funds. Come then to our help, and let us pray and labor together.

YOURS, AFFECTIONATELY, L.T.Y.

[1]This refers to the Reverend John R. McDowall, who first summoned New York women to moral reform. In 1836 he was tried and found guilty of ministerial misconduct and slander by the Third Presbytery of New York City — a presbytery made up largely of clergy and churches otherwise sympathetic to evangelical revival and reform movements. He died that year at age 35.

48

HENRY CLARKE WRIGHT

Marriage and Parentage

1858

Wright's new career as an itinerant lecturer and writer on family reform proved more successful than his earlier work with the antislavery movement. His support of women's rights shifted from advocating women's public rights in the 1830s to promoting women's rights to control their bodies in the 1850s.

A man has no right to compel his wife to lie or murder. He has no more right to compel her to yield to his passion, and thus to lie against the instincts of her nature, and kill the yearnings of her soul for true companionship, by urging upon her a passion which swallows up all other forces of her nature. No matter whether the violence that enables the man, under the name of husband, to enforce upon her the conditions of maternity, be in his own superior muscular energy, or in the shape of civil law or social and ecclesiastical sanction, the outrage upon her person is the same. . . .

Man can perpetrate no deeper wrong to himself, to his wife and child, and to his domestic peace, than to urge upon his wife maternity, when he knows her nature rebels against it. Nor can woman commit a greater crime against herself and her child, than to consent to become a mother, when her nature not only does not call for it, but actively repels it.

Let the wife say to the husband, "Show me thy love in some gentler way; let my head repose upon thee as upon a rock of trust; let me feel thine arms around me, to defend me from all harm, not to bring it to me." . . . Who but a ruffian would disregard such a request? Who but a being less than a man would say, "No matter how *you* feel, *I* wish to be gratified." The wife should be the regulator of *this* marriage relation, for only in obedience to the laws of her nature can she hope to continue to be the loving, healthful, happy wife. . . .

Henry C. Wright, *Marriage and Parentage or The Reproductive Element in Man, as a Means to His Elevation and Happiness* (Boston: Bela Marsh, 1858), 252–53, 301–02.

The man who regards the presence of the reproductive element in himself as a means of sensuous gratification, and marriage as a licensed mode of expending that element and of obtaining that gratification, can never hope to make for himself a pure and happy home. . . . By a constant expenditure of the vital element of his manhood, he enfeebles his reason, his conscience, his affection, and his power to love and appreciate his wife and child. He becomes repulsive, and incapable of forming true family relations. On our knowledge of the natural laws which should govern the expenditure of the reproductive element, and on our obedience to them, depend the question of a happy home.

EPILOGUE: THE NEW MOVEMENT SPLITS OVER THE QUESTION OF RACE, 1850–1869

49

JANE SWISSHELM

The Saturday Visiter

November 2, 1850

Editor of The Saturday Visiter, *the only antislavery paper in Pittsburgh from 1848 to 1857, Jane Swisshelm articulated the divisions over race that immediately began to appear within the emerging women's rights movement. In her 1850 debate with Parker Pillsbury she articulated a position that represented a distinct minority within the new movement. Arguing that women's rights should be separate from race, she accepted the inequalities between white and black women that the Grimkés had combatted.*

. . . At the time of writing we have only seen the first day's proceedings [of the Worcester Convention]. These are all we could have wished except the introduction of the color question. The convention was not

The Saturday Visiter, Nov. 2, 1850.

called to discuss the rights of color; and we think it was altogether irrelevant and unwise to introduce the question. We dislike very much the omnibus plan of action, and . . . we would contend to the last possible moment against any bundle of measures, even though we were in favor of every one taken separately and singly. In a woman's rights convention the question of color had no right to a hearing. One thing at a time! Always do one thing at a time, and you will get along much faster than by attempting to do a dozen. The question of the rights of colored men is already before the people. Let it work out its own salvation in its own strength. Many a man is in favor of emancipating every Southern slave, and granting the rights of citizenship to every free negro, who is by no means agreed that his wife or mother should stand on a political equality with himself. Many a man believes his wife and mother to be inferior to his bootblack, and many a woman ranks herself in the same scale. Then there are many of both sexes who are, or would be, anxious for the elevation of woman as such, who nevertheless hate "the niggers" most sovereignly. Why mingle the two questions? For our part we would say no resolution should be passed at that convention that would not have been as acceptable to the citizens of Georgia as to those of Massachusetts.

We are pretty nearly out of patience with the dogged perseverance with which so many of our reformers persist in their attempts to do everything at once. . . .

The subject of woman's admission to the rights of citizenship is of sufficient importance to claim consideration as a separate measure. . . . This convention was called to discuss Woman's rights, and if it had paid right good attention to its own business, it would have had work plenty.

50

PARKER PILLSBURY

Letter to Jane Swisshelm

November 18, 1850

In a letter to Swisshelm that was reprinted in* The North Star, *Pillsbury represented the well-established view of the Garrisonian movement and probably the majority within the women's convention movement that race had to be mentioned at women's rights conventions if those meetings were to encompass the needs of black women. Pillsbury, an editor and former minister, was among the most radical abolitionist leaders, especially in his condemnation of the pro-slavery sympathies of the northern clergy and in his support of women's rights within the antislavery movement.

DEAR MRS. SWISSHELM:

In the last Visiter, you say of a resolution relating to people of color, offered by Mr. Wendell Phillips in the late Convention of Women, at Worcester, Mass.—

"We are pretty nearly out of patience with the dogged perseverance with which so many of our Reformers persist in their attempt to do everything at once." And again: "In a Woman's Rights Convention, the question of color had no right to a hearing."

It seemed as though the usually kindly spirit and good judgment of the Visiter were a little wanting in these two utterances. I should not have noticed it at all in most of the public journals—indeed, I neither know nor care what but a few of them do say; for I should no more think of having them in my house, political or religious, than I would of inoculating the family with the foulest leprosy that ever unjointed the bones of a son of Abraham. But your Visiter finds ready entrance and cheerful greeting, so that we are a little solicitous about its bearing. . . .

But by way of explanation, (or if you please, apology), permit me to say that colored persons are held in such estimation in this country, that you must specify them whenever or wherever you mean to include them.

"Woman's Rights Convention and People of Color," *The North Star,* Dec. 5, 1850.

Lyceums, circuses, menageries, ballrooms, billiard-rooms, conventions, everything, "the Public are respectfully invited to attend." But who ever dreamed that "the public" meant anything colored? From church and theatre; from stage-coach, steam-ship and creeping canal-boat; from the infant school, law school and theological seminary; from museum, athenaeum and public garden, the colored race are either excluded altogether, or are admitted only by sufferance, or some very special arrangement, and under disadvantages to which no white person would or should submit for a moment. . . .

We have striven to separate the Ethiopian from all claim to human recognition and human sympathy. Nobody but abolitionists ever mean *colored* people, no matter how often they speak of "the public," or of their "fellow citizens" or "fellow sinners." . . .

And his race know it and feel it, as we cannot. Even the women's Convention demonstrated this, for scarcely a colored person, man or woman, appeared in it.

On the large committees appointed to carry out the plans of the Convention, embracing many persons in all, not a single colored member was placed. It is to be presumed that nobody thought of it, for we are not expected to think of colored people at all.

Under such circumstances, is it strange, is it an unpardonable sin, is it "dogged perseverance," to declare in a Convention called to demand and extend the rights of women, that we mean women of sable as well as sallow complexion? of the carved in ebony as well as the chiseled in ivory? If we did thus mean, the Convention should not have been held, or being held, it would only deserve the scorn and contempt of every friend of God and his children. Color was not discussed there — it need not have been. But it was heeded that the declaration be made in regard to it. That ANY woman have rights, will scarcely be believed; but that colored women have rights, would never have been thought of, without a specific declaration.

Most truly yours,
Parker Pillsbury
Concord, N. H., Nov. 18

51

JANE SWISSHELM

"Woman's Rights and the Color Question"

November 23, 1850

In her reply to Pillsbury, Swisshelm argued in favor of single-issue reform, and said the women's rights movement had no obligation to eradicate differences based on race or class.

We give [print] Mr. Pillsbury's article on this subject, and if we failed to prove the bad policy of linking these two questions, Mr. Pillsbury will surely succeed. Every thing he says about the exclusion of colored people from places and positions they have a right to occupy, is so much against uniting their cause to that of woman. The women of this glorious Republic are sufficiently oppressed without linking their cause to that of the slave. The slave is sufficiently oppressed without binding him to the stake which has ever held woman in a state of bondage. There is no kind of reason why the American prejudice against color should be invoked to sink woman into a lower degradation than that she already enjoys — no kind of reason why the car of emancipation, for the slave, should have been clogged by tying to its wheels the most unpopular reform that ever was broached, by having all the women in the world fastened to its axle as a drag. . . .

You, sir, show dogged perseverance in insisting that our starving seamstresses shall not strike for higher wages unless they put in a protest in favor of the boot-blacks — that woman shall not strike off her shackles until she can liberate every man that wears one — that she shall take no step forward until she overcomes a prejudice which oppresses another branch of humanity. You, Mr. Pillsbury, and the rest of your male coadjutors, enjoying all the rights for which women contend, have not been able to conquer the American prejudice against color, and now you expect that woman, crippled, helpless, bound, shall do what you have failed to perform with the free use of all your powers and faculties! . . .

The Saturday Visiter, Nov. 23, 1850.

As for colored women, all the interest they have in this reform is *as women*. All it can do for them is to raise them to the level of men of their own class. Then as that class rises let them rise with it. We only claim for a white wood-sawyer's wife that she is as good as a white wood-sawyer — a blacksmith's mother is a good as a blacksmith — a lawyer's sister is as good as a lawyer; at least this is our way of understanding this question. . . . The call [to the Convention] was explicit. It was to discuss the rights of Sex. We signed that call . . . and had no thought it was to be converted into an abolition meeting. With quite as much propriety it might have been turned into a Temperance or Law-Reform meeting, or a meeting to express sympathy with the Hungarian refugees. . . . We feel as if our name had been used for a purpose for which we did not give it, and we know of other signers of that call who are in the same predicament. It was a breach of trust, and one we shall remember when our name is asked for to answer another call.

52

FRANCES ELLEN WATKINS HARPER

Speech at the Eleventh Woman's Rights Convention

New York, May 1866

At this first women's rights convention after the Civil War, the prominent African American writer Frances Harper told how her support of women's rights was based on her own experience. But she also documented how the racism of American society differentiated the experience of black and white women, black men, and white men.

"We Are All Bound Up Together"

Mrs. Harper, of Ohio, said:

I feel I am something of a novice upon this platform. Born of a race whose inheritance has been of outrage and wrong, most of my life had been spent in battling against those wrongs. But I did not feel as keenly

Proceedings of the Eleventh Woman's Rights Convention, May, 1866 (New York: Robert Johnston, 1866), 45–48.

as others, that I had these rights, in common with other women, which are now demanded. About two years ago, I stood within the shadows of my home. A great sorrow had fallen upon my life. My husband had died suddenly, leaving me a widow, with four children, one my own, and the others step-children. I tried to keep my children together. But my husband died in debt; and before he had been in his grave three months, the administrator had swept the very milk-crocks and wash tubs from my hands. I was a farmer's wife and made butter for the Columbus market; but what could I do, when they had swept all away? They left me one thing — and that was a looking-glass! Had I died instead of my husband, how different would have been the result! By this time he would have had another wife, it is likely; and no administrator would have gone into his house, broken up his home, and sold his bed, and taken away his means of support.

. . . And I went back to Ohio with my orphan children in my arms, without a single feather bed in this wide world, that was not the custody of the law. I say, then, that justice is not fulfilled so long as woman is unequal before the law.

We are all bound up together in one great bundle of humanity, and society cannot trample on the weakest and feeblest of its members without receiving the curse in its own soul. You tried that in the case of the negro. You pressed him down for two centuries; and in so doing you crippled the moral strength and paralyzed the spiritual energies of the white men of the country. When the hands of the black were fettered, white men were deprived of the liberty of speech and the freedom of the press. Society cannot afford to neglect the enlightenment of any class of its members. At the South, the legislation of the country was in behalf of the rich slaveholders, while the poor white man was neglected. What is the consequence to-day? From that very class of neglected poor white men, comes the man who stands to-day with his hand upon the helm of the nation. He fails to catch the watchword of the hour, and throws himself, the incarnation of meanness, across the pathway of a nation. My objection to [President] Andrew Johnson is not that he has been a poor white man; my objection is that he keeps "poor whites" all the way through. (Applause.) That is the trouble with him.

This grand and glorious revolution which has commenced, will fail to reach its climax of success, until throughout the length and breadth of the American Republic, the nation shall be so color-blind, as to know no man by the color of his skin or the curl of his hair. It will then have no privileged class, trampling upon and outraging the unprivileged classes, but will be then one great privileged nation, whose privilege will be to produce the loftiest manhood and womanhood that humanity can attain.

I do not believe that giving the woman the ballot is immediately going to cure all the ills of life. I do not believe that white women are dewdrops just exhaled from the skies. I think that like men they may be divided into three classes, the good, the bad, and the indifferent. The good would vote according to their convictions and principles; the bad, as dictated by prejudice or malice; and the indifferent will vote on the strongest side of the question, with the winning party.

You white women speak here of rights. I speak of wrongs. I, as a colored woman, have had in this country an education which has made me feel as if I were in the situation of Ishmael, my hand against every man, and every man's hand against me.[1] Let me go to-morrow morning and take my seat in one of your street cars — I do not know that they will do it in New York, but they will in Philadelphia — and the conductor will put up his hand and stop the car rather than let me ride.

A lady: They will not do that here.

Mrs. Harper: They do in Philadelphia. Going from Washington to Baltimore this Spring, they put me in the smoking car. (Loud voices — "Shame.") Aye, in the capital of the nation, where that black man consecrated himself to the nation's defence, faithful when the white man was faithless, they put me in the smoking car! They did it once; but the next time they tried it, they failed; for I would not go in. I felt the fight in me; but I don't want to have to fight all the time. To-day I am puzzled where to make my home. I would like to make it in Philadelphia, near my own friends and relations. But if I want to ride in the streets of Philadelphia, they send me to ride on the platform with the driver. (Cries of "Shame.") Have women nothing to do with this? Not long since, a colored woman took her seat in an Eleventh street car in Philadelphia, and the conductor stopped the car, and told the rest of the passengers to get out, and left the car with her in it alone, when they took it back to the station. One day I took my seat in a car, and the conductor came to me and told me to take another seat, I just screamed "murder." The man said if I was black I ought to behave myself. I knew that if he was white he was not behaving himself. Are there not wrongs to be righted?

In advocating the cause of the colored man, since the Dred Scott decision, I have sometimes said that I thought the nation had touched bottom. But let me tell you there is a depth of infamy lower than that. It is when the nation, standing upon the threshold of a great peril, reached out its hands to a feebler race, and asked that race to help it and when the peril was over, said, You are good enough for soldiers, but not good

[1] Genesis 16:12.

enough for citizens. When Judge Taney said that the men of my race had no rights which the white man was bound to respect, he had not seen the bones of the black man bleaching outside of Richmond.[2] He had not seen the thinned ranks and the thickened graves of the Louisiana Second, a regiment which went into battle nine hundred strong, and came out with three hundred. He had not stood at Olustee and seen defeat and disaster crushing down the pride of our banner, until word was brought to Col. Hallowell, "The day is lost; go in and save it;" and black men stood in the gap, beat back the enemy, and saved your army. (Applause.)

We have a woman in our country who has received the name of "Moses," not by lying about it, but by acting it out (applause) — a woman who has gone down into the Egypt of slavery and brought out hundreds of our people into liberty.[3] The last time I saw that woman, her hands were swollen. That woman who had led one of Montgomery's most successful expeditions, who was brave enough and secretive enough to act as a scout for the American army, had her hands all swollen from a conflict with a brutal conductor, who undertook to eject her from her place. That woman, whose courage and bravery won a recognition from our army and from every thoroughfare of travel. Talk of giving women the ballot-box? Go on. It is a normal school, and the white women of the country need it. While there exists this brutal element in society which tramples upon the feeble and treads down the weak, I tell you that if there is any class of people who need to be lifted out of their airy nothings and selfishness, it is the white women of America. (Applause.)

[2] In the 1857 Dred Scott case, Roger B. Taney (1777–1864), fifth chief justice of the U.S. Supreme Court, writing for the majority, held that free blacks could not possess rights of citizenship entitling them to sue in federal courts; that Congress had no power to exclude slavery from the territories, as provided for in the Missouri Compromise; and that slaves who escaped to the North could be legally recaptured and returned to their owners in the South.

[3] Refers to Harriet Tubman (c. 1820–1913), a fugitive slave who in the 1850s returned to Maryland around nineteen times and guided to freedom between sixty and three hundred slaves.

53

EQUAL RIGHTS ASSOCIATION

Proceedings

New York City, May 1869

Created at the 1866 Woman's Rights Convention in New York to support both black suffrage and woman suffrage, the American Equal Rights Association offered a timely forum for a renewal of the cross-race coalition of the 1830s. However, debate at the convention reflected the formidable obstacles that blocked such a coalition.

Mr. [*Frederick*] *Douglass:* I must say that I do not see how any one can pretend that there is the same urgency in giving the ballot to woman as to the negro. With us, the matter is a question of life and death, at least, in fifteen States of the Union. When women, because they are women, are hunted down through the cities of New York and New Orleans, when they are dragged from their houses and hung upon lamp-posts; when their children are torn from their arms, and their brains dashed out upon the pavement; when they are objects of insult and outrage at every turn; when their children are not allowed to enter schools; then they will have an urgency to obtain the ballot equal to our own. (Great applause.)

A Voice: Is that not all true about black women?

Mr. Douglass: Yes, yes, yes; it is true of the black woman, but not because she is a woman, but because she is black. (Applause.) Julia Ward Howe at the conclusion of her great speech delivered at the convention in Boston last year, said: "I am willing that the negro shall get the ballot before me." (Applause.) Woman! why, she has 10,000 modes of grappling with her difficulties. I believe that all the virtue of the world can take care of all the evil. I believe that all the intelligence can take care of all the ignorance. (Applause.) I am in favor of woman's suffrage in order that we shall have all the virtue and vice confronted. Let me

---

Elizabeth Cady Stanton, Susan B. Anthony, and Matilda Joslyn Gage, eds., *History of Woman Suffrage* (New York: Fowler & Wells, 1882), 2:382, 391–92, 397; and Proceedings, American Equal Rights Association, New York City, 1869; *Stanton-Anthony Papers,* reel 13, frame 0504.

tell you that when there were few houses in which the black man could have put his head, this woolly head of mine found a refuge in the house of Mrs. Elizabeth Cady Stanton, and if I had been blacker than sixteen midnights, without a single star, it would have been the same. (Applause.)

Miss [*Susan B.*] *Anthony:* The old anti-slavery school says women must stand back and wait until the negroes shall be recognized. But we say, if you will not give the whole loaf of suffrage to the entire people, give it to the most intelligent first. (Applause.) If intelligence, justice, and morality are to have precedence in the Government, let the question of woman be brought up first and that of the negro last. (Applause.) While I was canvassing the State with petitions and had them filled with names for our cause to the Legislature, a man dared to say to me that the freedom of women was all a theory and not a practical thing. (Applause.) When Mr. Douglass mentioned the black man first and the woman last, if he had noticed he would have seen that it was the men that clapped and not the women. There is not the woman born who desires to eat the bread of dependence, no matter whether it be from the hand of father, husband, or brother; for any one who does so eat her bread places herself in the power of the person from whom she takes it. (Applause.) Mr. Douglass talks about the wrongs of the negro; but with all the outrages that he to-day suffers, he would not exchange his sex and take the place of Elizabeth Cady Stanton. (Laughter and applause.) . . .

Mrs. Lucy Stone: Mrs. Stanton will, of course, advocate the precedence for her sex, and Mr. Douglass will strive for the first position for his, and both are perhaps right. If it be true that the government derives its authority from the consent of the governed, we are safe in trusting that principle to the uttermost. If one has a right to say that you can not read and therefore can not vote, then it may be said that you are a woman and therefore can not vote. We are lost if we turn away from the middle principle and argue for one class. . . . Over in New Jersey they have a law which says that *any* father—he might be the most brutal man that ever existed—*any* father, it says, whether he be under age or not, may by his last will and testament dispose of the custody of his child, born or to be born, and that such disposition shall be good against all persons, and that the mother may not recover her child; and that law modified in form exists over every State in the Union except Kansas. Woman has an ocean of wrongs too deep for any plummet, and the negro, too, has an ocean of wrongs that can not be fathomed. There are two great oceans; in one is the black man, and in the other is the woman. But I thank God for that XV Amendment, and hope that it will be adopted in every State. I will be thankful in my soul if *any* body can get out of the terrible pit. But I believe that the safety of the government would be more promoted by the admission of woman as an element of restoration and harmony than the negro. I believe that the influence of woman will save the country before every other power. (Applause.) I see the signs of the times pointing to this consummation, and I believe that in some parts of the country women will vote for the President of these United States in 1872. (Applause.) . . .

Mrs. Paulina W. Davis said that she would not be altogether satisfied to have the XVth Amendment passed without the XVIth, for woman have a race of tyrants raised above her in the South, and the black women of that country would also receive worse treatment than if the Amendment was not passed.[1] . . .

Mr. Douglass said that all the disinterested spectators would concede that this Equal Rights meeting had been pre-eminently a Woman's Rights meeting. (Applause.) They had just heard an argument with which he could not agree — that the suffrage to the black men should be postponed to that of the women. . . .

The President, Mrs. Stanton, argued that not another man should be enfranchised until enough women are admitted to the polls to outweigh those already there. (Applause.) She did not believe in allowing ignorant negroes and foreigners to make laws for her to obey. (Applause.)

Mrs. Harper (colored) asked Mr. Blackwell to read the fifth resolution of the series he submitted, and contended that that covered the whole ground of the resolutions of Mr. Douglass. When it was a question of race, she let the lesser question of sex go. But the white women all go for sex, letting race occupy a minor position. She liked the idea of working women, but she would like to know if it was broad enough to take colored women?

[Blackwell's fifth proposed resolution: "Resolved, That any party professing to be democratic in spirit or republican in principle, which opposes or ignores the political rights of woman, is false to its professions, short sighted in its policy, and unworthy of the confidence of the friends of impartial liberty."]

Miss Anthony and several others: Yes, yes.

Mrs. Harper said that when she was at Boston there were sixty women who left work because one colored woman went to gain a liveli-

[1] Davis was hoping that the next Amendment would provide for women's suffrage.

hood in their midst. (Applause.) If the nation could only handle one question, she would not have the black women put a single straw in the way, if only the men of the race would obtain what they wanted. (Great applause.) . . .

Miss Anthony protested against the XVth Amendment because it wasn't Equal Rights. It put two million more men in position of tyrants over two million women who had until now been the equals of the men at their side. . . .

54

Founding of the National Woman Suffrage Association

New York, 1869

Two days after the Equal Rights Association adjourned, Stanton and Anthony held a reception that became the founding meeting of the National Woman Suffrage Association. The association's goal was the passage of a sixteenth amendment to guarantee woman suffrage. That year Lucy Stone and other supporters of the Fifteenth Amendment organized the American Woman Suffrage Association in Boston. The two groups finally merged in 1890. Another thirty years of organizing were required to achieve the ratification of the Nineteenth Amendment, which guaranteed women's right to vote in 1920.

Out of these broad differences of opinion on the amendments, as shown in the debates, divisions grew up between Republicans and Abolitionists on the one side, and the leaders of the Woman Suffrage movement on the other. The constant conflict on the Equal Rights platform proved the futility of any attempt to discuss the wrongs of different classes in one association. A general dissatisfaction had been expressed by the delegates from the West at the latitude of debate involved in an Equal Rights Association. Hence, a change of name and more restricted discussions were strenuously urged by them. Accordingly . . . a meeting was called . . . which resulted in reorganization under the name of "The National Woman Suffrage Association." . . .

Stanton et al., *History of Woman Suffrage,* 2:400.

Delegates from nineteen States, including California and Washington Territory, were present on the occasion, and all felt the importance of an organization distinctively for Woman's Suffrage, in view of the fact that a Sixteenth Amendment to the Federal Constitution to secure this is now before the people. The Association has held several meetings to plan the work for the coming year. Committees are in correspondence with friends in the several States to complete the list of officers.

Questions for Consideration

1. In what kinds of activities did antislavery women engage?
2. What were the goals of the petition campaigns?
3. How did the antislavery movement challenge established notions of manhood and womanhood?
4. Why was the issue of racial prejudice linked to the goal of immediate abolition?
5. What were the sources of the Grimké sisters' opposition to slavery?
6. Why did Angelina and Sarah Grimké defy convention and advocate women's rights?
7. What were the Grimkés trying to achieve by speaking out on women's rights? Were they successful?
8. Why did some of the clergy become their main opponents? On what did they base their opposition to emancipation and women's rights?
9. Why did the Grimkés claim women's equality in church governance? Why in secular governance? On what did they base their claims?
10. What was "non-resistance" and why did Angelina Grimké support it?
11. Why was Lucretia Mott the center of so much attention at the 1840 antislavery convention in London?
12. Who organized the 1848 Seneca Falls Women's Rights Convention, and why? What did they accomplish there?
13. How did the 1848 convention differ from the 1837 women's antislavery convention?
14. Was the women's rights convention movement of the 1850s successful? Why or why not?
15. Why did the emerging woman suffrage movement split on the issue of race? What effect did the split have on these two issues?

Selected Bibliography

PRIMARY SOURCES

Barnes, Gilbert H., and Dwight L. Dumond, eds. *Letters of Theodore Dwight Weld, Angelina Grimké Weld and Sarah Grimké, 1822–1844,* 2 vols. New York: Appleton-Century-Crofts, 1934; reprinted Gloucester, Mass.: Smith, 1965.

Ceplair, Larry. *The Public Years of Sarah and Angelina Grimké: Selected Writings, 1835–1839.* New York: Columbia University Press, 1989.

DuBois, Ellen Carol, ed. *Elizabeth Cady Stanton, Susan B. Anthony: Correspondence, Writings, Speeches.* New York: Schocken, 1981.

Foner, Philip S., ed. *Frederick Douglass on Women's Rights.* Westport, Conn.: Greenwood Press, 1976.

Foster, Frances Smith, ed. *A Brighter Coming Day: A Frances Ellen Watkins Harper Reader.* New York: Feminist Press, 1990.

Gordon, Ann D., ed. *The Selected Papers of Elizabeth Cady Stanton and Susan B. Anthony: Volume 1, In the School of Anti-Slavery, 1840–1866.* New Brunswick: Rutgers University Press, 1997.

Greene, Dana, ed. *Lucretia Mott: Her Complete Speeches and Sermons.* New York: Mellen, 1980.

Hallowell, Anna Davis, ed. *James and Lucretia Mott: Life and Letters.* Boston: Houghton Mifflin, 1884.

McClymer, John F. *This High and Holy Moment: The First National Woman's Rights Convention, Worcester, 1850.* New York: Harcourt Brace, 1999.

Meltzer, Milton, and Patricia G. Holland, eds. *Lydia Maria Child: Selected Letters, 1817–1880.* Amherst: University of Massachusetts Press, 1982.

Richardson, Marilyn, ed., *Maria W. Stewart, America's First Black Woman Political Writer: Essays and Speeches.* Bloomington: Indiana University Press, 1987.

Turning the World Upside Down: The Anti-Slavery Convention of American Women, Held in New York City, May 9–12, 1837. New York: Feminist Press, 1987.

Wright, Daniel. "What Was the Appeal of Moral Reform to Antebellum Northern Women?" in "Women and Social Movements in the United States, 1830–1930," a Web site at http://womhist.binghamton.edu.

WOMEN ABOLITIONISTS AND WOMEN'S RIGHTS

Bacon, Margaret Hope. *Valiant Friend: The Life of Lucretia Mott.* New York: Walker, 1980.

Barry, Kathleen. *Susan B. Anthony: A Biography.* New York: New York University Press, 1988.

Birney, Catherine H. *The Grimké Sisters: Sarah and Angelina Grimké: The First American Women Advocates of Abolition and Women's Rights.* Philadelphia: Lee and Shepard, 1885.

DuBois, Ellen Carol. *Feminism and Suffrage: The Emergence of an Independent Women's Movement in America, 1848–1869.* Ithaca: Cornell University Press, 1978.

Gordon, Ann D., with Bettye Collier-Thomas, John H. Bracey, Arlene Voski Avakian, and Joyce Avrech Berkman, ed., *African American Women and the Vote, 1837–1965.* Amherst: University of Massachusetts Press, 1997.

Griffith, Elisabeth. *In Her Own Right: The Life of Elizabeth Cady Stanton.* New York: Oxford University Press, 1984.

Hansen, Debra Gold. *Strained Sisterhood: Gender and Class in the Boston Female Anti-Slavery Society.* Amherst: University of Massachusetts Press, 1993.

Hardesty, Nancy A. *Your Daughters Shall Prophesy: Revivalism and Feminism in the Age of Finney.* Brooklyn: Carlson, 1991.

Hersh, Blanche Glassman. *The Slavery of Sex: Feminist-Abolitionists in America.* Urbana: University of Illinois Press, 1978.

Hewitt, Nancy A. *Women's Activism and Social Change: Rochester, New York, 1822–1872.* Ithaca: Cornell University Press, 1984.

Isenberg, Nancy. *Sex and Citizenship in Antebellum America.* Chapel Hill: University of North Carolina Press, 1998.

Kraditor, Aileen S. *Means and Ends in American Abolitionism: Garrison and His Critics on Strategy and Tactics, 1834–1850.* New York: Random House, 1967.

Jeffrey, Julie Roy. *The Great Silent Army of Abolitionism: Ordinary Women in the Antislavery Movement.* Chapel Hill: University of North Carolina Press, 1998.

Lerner, Gerda. *The Grimké Sisters of North Carolina: Pioneers for Women's Rights and Abolitionism.* New York: Oxford University Press, 1967.

———. *The Feminist Thought of Sarah Grimké.* New York: Oxford University Press, 1998.

———. "The Grimké Sisters and the Struggle against Race Prejudice," *Journal of Negro History* 26 (October 1963): 277–91.

Lumpkin, Katharine DePre. *The Emancipation of Angelina Grimké.* Chapel Hill: University of North Carolina Press, 1974.

Mabee, Carleton, with Susan Mabee Newhouse. *Sojourner Truth: Slave, Prophet, Legend.* New York: New York University Press, 1993.

McFadden, Margaret H. *Golden Cables of Sympathy: The Transatlantic Sources of Nineteenth-Century Feminism.* Lexington: University of Kentucky Press, 1999.
McKivigan, John R., ed. *The American Abolitionist Movement: A Collection of Scholarly Articles Illustrating Its History. Vol. 4: Abolitionism and Issues of Race and Gender.* Hamden, Conn.: Garland, 1999.
Melder, Keith E. *Beginnings of Sisterhood: The American Woman's Rights Movement, 1800–1850.* New York: Schocken, 1977.
Painter, Nell Irvin. *Sojourner Truth: A Life, A Symbol.* New York: Norton, 1996.
Peterson, Carla L. *"Doers of the Word": African American Women Speakers and Writers in the North (1830–1880).* New York: Oxford University Press, 1995.
Ryan, Mary P. *Women in Public: Between Banners and Ballots, 1825–1880.* Baltimore: Johns Hopkins University Press, 1990.
Sklar, Kathryn Kish. *Catharine Beecher: A Study in American Domesticity.* New Haven: Yale University Press, 1973.
Stansell, Christine. "Woman in Nineteenth-Century America." *Gender and History* 11, no. 3 (November 1999): 419–32.
Terborg-Penn, Rosalyn. *African-American Women in the Struggle for the Vote, 1850–1920.* Bloomington: Indiana University Press, 1998.
Yee, Shirley J. *Black Women Abolitionists: A Study in Activism, 1828–1860.* Knoxville: University of Tennessee Press, 1992.
Yellin, Jean Fagan. *Women and Sisters: The Antislavery Feminists in American Culture.* New Haven: Yale University Press, 1989.
Yellin, Jean Fagan, and John C. Van Horne, eds. *The Abolitionist Sisterhood: Women's Political Culture in Antebellum America.* Ithaca: Cornell University Press, 1994.

RELIGION AND THE ANTISLAVERY MOVEMENT

Abzug, Robert H. *Passionate Liberator: Theodore Dwight Weld and the Dilemma of Reform.* New York: Oxford University Press, 1980.
———. *Cosmos Crumbling: American Reform and the Religious Imagination.* New York: Oxford University Press, 1994.
Goodman, Paul. *Of One Blood: Abolitionism and the Origins of Racial Equality.* Berkeley: University of California Press, 1998.
Hatch, Nathan O. *The Democratization of American Christianity.* New Haven: Yale University Press, 1989.
McKivigan, John R. and Mitchell Snay, eds. *Religion and the Antebellum Debate over Slavery.* Athens: University of Georgia Press, 1998.
Perry, Lewis. *Radical Abolitionism: Anarchy and the Government of God in Antislavery Thought.* Knoxville: University of Tennessee Press, 1995.

Index

Printed in the United States
By Bookmasters